FALLACIES

by the same author

ELEMENTARY FORMAL LOGIC

Fallacies

C. L. HAMBLIN

*Professor of Philosophy
University of New South Wales*

METHUEN & CO LTD

First published in 1970
by Methuen & Co Ltd
11 New Fetter Lane, London EC4
© 1970 C. L. Hamblin
Printed in Great Britain by
Richard Clay (The Chaucer Press) Ltd
Bungay, Suffolk

SBN 416 14570 1

Distributed in the USA
by Barnes & Noble Inc

CONTENTS

	ACKNOWLEDGEMENTS	7
1.	The Standard Treatment	9
2.	Aristotle's List	50
3.	The Aristotelian Tradition	89
4.	Arguments 'Ad'	135
5.	The Indian Tradition	177
6.	Formal Fallacies	190
7.	The Concept of Argument	224
8.	Formal Dialectic	253
9.	Equivocation	283
	BIBLIOGRAPHY	304
	INDEX	317

ACKNOWLEDGMENTS

I have picked many colleagues' brains, but most of them must here be thanked anonymously. I am especially indebted to Professor L. M. De Rijk for the loan of a vital microfilm; to D. D. McGibbon for help with Greek texts; to Fr Romuald Green for making available his work on Obligation (which I wish he would get round to publishing); and to Professor Nicholas Rescher for remedying part of my deep ignorance of the Arabs. It should be added that A. N. Prior started the whole thing off by getting me interested in Buridan.

Dedication? To the friend who said 'I hope the title isn't an accurate description of the contents'. But most of all to Rita, Fiona and Julie.

C.L.H.

CHAPTER I

The Standard Treatment

There is hardly a subject that dies harder or has changed so little over the years. After two millennia of active study of logic and, in particular, after over half of that most iconoclastic of centuries, the twentieth A.D., we still find fallacies classified, presented and studied in much the same old way. Aristotle's principal list of thirteen types of fallacy in his *Sophistical Refutations* – the Latin title is *De Sophisticis Elenchis* (from Greek Περὶ Τῶν Σοφιστικῶν Ἐλέγχων) whence they have often been called 'sophisms', and sometimes 'elenchs' – still appears, usually with one or two omissions and a handful of additions, in many modern textbooks of logic; and though there have been many proposals for reform, none has met more than temporary acceptance. Such set-backs as Aristotle's treatment has had have been as much due to irrelevant vicissitudes of history as to any kind of criticisms of its shortcomings. Thus, although current in the ancient world in Athens, Alexandria and Rome it was 'lost' to western Europe, for some centuries during the monastic period; but was rediscovered with enthusiasm about the twelfth century, when it began to form a section of the logic curriculum in the emerging universities. Since that time until the present century textbooks of logic not containing a short chapter on fallacies have been the exception; and since, for most of the period, all students took Logic, Europe's men-of-affairs have generally regarded a nodding acquaintance with a standard version of Aristotle's doctrine as a routine necessity of the same character as knowledge of the multiplication table. Quite a few of these men, in fact, have written accounts of fallacies themselves; they include at least one Pope, two saints, archbishops in profusion, the first Chancellor of the University of Oxford, and a Lord Chancellor of England. The tradition has repeatedly proved too strong for its dissentients. Ramus, in the sixteenth century, led an attack on Aristotle and refused to consider fallacies as a proper subject for Logic on the grounds that the study of *correct* reasoning was enough in itself to make their nature clear; but within a few years his own followers had reinstated the subject and one of them, Heizo Buscher, actually published a treatise entitled *The Theory of the Solution of Fallacies . . . deduced and explained from the logic of P. Ramus*.[1] Bacon and Locke also dropped the Aristotelian treatment, but only to replace it with treatments of their own which, in due course, became partially fused with it again. During the past century some of the more mathematically minded of logicians, starting with Boole, have dropped the subject from their books in apparent agreement with Ramus; but it is possible to discern a trend back.

What about other traditions than our own? Constantinople, in the interval between the decline of Rome and its own fall to the Turks, continued the Greek tradition that was in decline further west; and some Arab logicians also inherited Aristotle's *Sophistical Refutations* and wrote their own commentaries on it. But these traditions were mere outposts of our own. Further east, we find an apparently independent logical tradition in India which, starting with the *Nyāya sūtra*, has its own doctrine of fallacies as an adjunct to its own theory of inference. Indian logicians have displayed the same concern to explore the forms of faulty reasoning, and the same inability either to move outside their original tradition or to dispense with it. The study of the Indian tradition is of especial interest here in providing us with a control on which to test our woollier historical generalizations.

Strangely, in a certain sense, there has never yet been a book on fallacies; never, that is, a book-length study of the subject as a whole, or of incorrect reasoning in its own right rather than as an afterthought or adjunct to something else. Schopenhauer's *Art of Controversy* is too short, and Bentham's *Book of Fallacies* too specialized, to qualify. A book entitled *Fallacies: a View of Logic from the Practical Side*, by Alfred Sidgwick, belies its title and is in large part concerned with a particular theory of non-fallacious logical reasoning. The medieval treatises, though some of them

[1] Buscherus, *De ratione solvendi sophismata* (3rd edn. 1594).

run to enormous length – that of St Albert the Great, for example, has 90,000 Latin words – are mere commentaries on Aristotle even when, as in the case of Peter of Spain's *Treatise on the Major Fallacies*, they do not indicate the fact in their titles. And all the others, including the wordy treatment by J. S. Mill, must be counted as short treatments in longer works. (Mill is just as wordy in the rest of his volume.) Even Aristotle's *Sophistical Refutations* is properly only the ninth book of his *Topics*.

There are, of course, works on fallacy of a slightly different kind; namely, less formal works such as Thouless's *Straight and Crooked Thinking*, Stebbing's *Thinking to Some Purpose* and, perhaps, Kamiat's *Critique of Poor Reason* and Stuart Chase's *Tyranny of Words* and *Guides to Straight Thinking*, which aim to induce in the reader an appreciation of and feeling for faulty reasoning by giving discussions based mainly on examples. Some of these books – I am not going to say which – are good, but they do not supply the need for a critical theoretical survey. Into the same category – or, perhaps, into the space between the two stools – I consign a book entitled *Fallacy – the Counterfeit of Argument*, by W. Ward Fearnside and William B. Holther. This is described on the back cover as '51 fallacies named, explained and illustrated'. The gratifyingly large bag of fallacies has been arranged in a system of categories partly resembling the traditional ones but not, it is to be presumed, intended either as exhaustive or as non-overlapping. These books have their place; but their place is not here. What is needed, above all, is discussion of some unresolved theoretical questions, which these books do not include in their terms of reference.

The truth is that nobody, these days, is particularly satisfied with this corner of logic. The traditional treatment is too unsystematic for modern tastes. Yet to dispense with it, as some writers do, is to leave a gap that no one knows how to fill. We have no *theory* of fallacy at all, in the sense in which we have theories of correct reasoning or inference. Yet we feel the need to ticket and tabulate certain kinds of fallacious inference-process which introduce considerations falling outside the other topics in our logic-books. In some respects, as I shall argue later, we are in the position of the medieval logicians before the twelfth century: we have lost the doctrine of fallacy, and need to rediscover it. But it is all more complicated than that because, these days, we set ourselves higher standards of theoretical rigour and will not be satisfied for long with a theory less ramified and systematic than we are used to in other departments of Logic; and one of the things we may find is that the kind of theory we need cannot be constructed in isolation from them. What I shall suggest is that interest in fallacies has always been, in part, misplaced in that the function of their study has been to remind the student (and his teacher) of features of the scope and limitations of the other parts of Logic. What the logicians of the thirteenth and fourteenth centuries made of the study of fallacies is especially interesting in this connection.

This is, however, for later chapters. To start with, let us set the stage with an account not of what went on in the thirteenth century, or even of what Aristotle wrote, but of the typical or average account as it appears in the typical short chapter or appendix of the average modern textbook. And what we find in most cases, I think it should be admitted, is as debased, worn-out and dogmatic a treatment as could be imagined – incredibly tradition-bound, yet lacking in logic and in historical sense alike, and almost without connection to anything else in modern Logic at all. This is the part of his book in which a writer throws away logic and keeps his reader's attention, if at all, only by retailing the traditional puns, anecdotes, and witless examples of his forbears. 'Everything that runs has feet; the river runs: therefore, the river has feet' – this is a medieval example, but the modern ones are no better. As a whole, the field has a certain fascination for the connoisseur, but that is the best that can be said for it.

A fallacious argument, as almost every account from Aristotle onwards tells you, is one that *seems to be valid* but *is not* so. Two different ways of classifying fallacies immediately present themselves. First, taking for granted that we have arguments that seem to be valid, we can classify them according to what it is that makes them not so; or secondly, taking for granted that they are not valid, we can classify them according to what it is that makes them seem to be valid. Most accounts take neither of these easy courses. Aristotle's original classification tries to be both sorts at once, and there are writers even in modern times who adopt it without criticism. Of those who invent their own classifications many share this uncertainty of purpose; and, in any case, their most noteworthy characteristic is that they disagree not only with the Aristotelians but also extensively with one another, and have quite failed to establish any account for longer than the time it takes a book to go out of print. In fact, though everyone has his classification, it is commonly argued that it is impossible to classify fallacies at all. De Morgan (*Formal Logic*, p. 276) writes:

> There *is* no such thing as a classification of the ways in which men may arrive at an error: it is much to be doubted whether there ever *can be*.

and Joseph (*Introduction to Logic*, p. 569) says

> Truth may have its norms, but error is infinite in its aberrations, and they cannot be digested in any classification.

but even they frequently express doubts. Cohen and Nagel (*Introduction to Logic and Scientific Method*, p. 382) say

> It would be impossible to enumerate all the abuses of logical principles occurring in the diverse matters in which men are interested.

They go on to consider 'certain outstanding abuses'.

Despite divergences of arrangement, there is a considerable overlap in raw material as between one writer and another: the individual kinds of fallacy are much the same, even down to their names. It will suit us, therefore, to forget about arrangement and describe the raw material. I shall start by running through the traditional list, and then discuss some additions. I am mainly concerned with recent accounts[1] but draw here and there on older ones.

EQUIVOCATION

Aristotle classified fallacies into those Dependent on Language and those Outside Language. (The traditional Latin terms are

[1] The recent books that I have especially consulted are: Cohen and Nagel, *Introduction to Logic and Scientific Method*; Black, *Critical Thinking*; Oesterle, *Logic: The Art of Defining and Reasoning*; Schipper and Schuh, *A First Course in Modern Logic*; Copi, *Introduction to Logic*; Salmon, *Logic*. Two dozen others could have been included. Oesterle is a strict traditionalist and the others all partly invent their own classifications.

in dictione and *extra dictionem*, from the Greek παρὰ τὴν λέξιν and ἔξω τῆς λέξεως.) Fallacies of the first category are those that arise from ambiguity in the words or sentences in which they are expressed. Those of the second category will occupy us later.

In the simplest case of Fallacies Dependent on Language the ambiguity can be traced to double-meaning in a single word. This is the Fallacy of Equivocation.

The word 'equivocation' refers literally to pairs of words that are the same in pronunciation. Ralph Lever,[1] one of the earliest writers on logic in English, translated *aequivoca* as 'lykesounding wordes', and its opposite *univoca* as 'playnmeaning wordes'. The term commonly has a pejorative sense, in that an equivocal argument is one *deliberately intended* to deceive; though, in spite of a distinction made by Max Black, this is not usually a part of the logician's meaning. At its lowest level Equivocation is plain punning: at least three modern American books I have consulted think it worth while to give the example 'Some dogs have fuzzy ears; my dog has fuzzy ears; therefore my dog is some dog!'; and Oesterle is only graver, not more in earnest, in quoting the traditional 'Whatever is immaterial is unimportant; whatever is spiritual is immaterial; therefore, whatever is spiritual is unimportant'. One of Abraham Fraunce's examples in Elizabethan times (*Lawier's Logike*, f. 27) was:

> All the maydes in Camberwell may daunce in an egge shell.

He explains:

> Of a little village by *London*, where *Camberwell* may be taken for the Well in the towne, or ye towne it selfe.

And again:

> So lastly, the Mayre of *Earith*, is the best Mayre next to the Mayre of *London*. Where the towne, God knowes, is a poore thing, and the Mayre thereof a seely fellow, in respect of the Mayres of divers other cities, yet it is the very next to *London*, because there is none betweene.

These examples serve to introduce us to different kinds of ambiguity. They do not, however, provide good examples of fallacies since, whatever our feelings about maids in Camberwell

---

[1] *The Arte of Reason, rightly termed Witcraft* (1573), pp. 2–3.

or the Mayor of Erith, we are hardly capable of being deceived by any serious chain of reasoning exploiting the double-meanings in the statements about them.

If we try to find better examples we meet another kind of difficulty, in that what is non-trivial may be controversial. Joseph attempts to illustrate Equivocation with discussion of an example as follows (p. 579):

> 'A mistake in point of law,' says Blackstone, 'which every person of discretion not only may, but is bound and presumed to know, is in criminal cases no sort of defence'; the State must perhaps presume a knowledge of the law, and so far we are bound to know it, in the sense of being required under penalty; but a criminal action done in ignorance of the law that a man is *legally* bound to know is often considered *morally* discreditable, as if the knowledge of the law on the matter were a plain moral duty. How far that is so in a particular case may be a very doubtful question; the maxim quoted tends to confuse the moral with the legal obligation.

All that Joseph claims, however, is that it is doubtful that moral and legal duty must be identified; not that it is clear that they must be distinguished. If moral words were not slippery there would be little need for the study of moral philosophy. For this to be a clear example of Equivocation there needs to be a clear distinction between moral rectitude and obedience to the law of the land. We know, of course, that there is *sometimes* a case for saying that the law is wrong and should be altered or even disobeyed. But the law of the land is interpreted by the courts, and the courts are inevitably and properly influenced to some extent by moral factors; and, on the other hand, it could be argued that a certain conformity to law, in so far as it promotes the general good through stable government, is morally commendable on its own account. We do not need to resolve these questions here, but it must be clear that there is at least room for debate. In many contexts the two subsenses of moral words can be conflated without risk, so that a charge of Equivocation needs to be backed up with a demonstration that the context is one in which the distinction is necessary.

The more satisfactory accounts of Equivocation are those which – usually at some length – give us hints and practice in looking for those slight shifts of direction which may lay a detailed argument open to objection. Max Black, for example, discusses four types of meaning shift which he calls 'Sign: Referent', 'Dictionary meaning: Contextual meaning', 'Connotation: Denotation' and 'Process: Product'. (See his Chapter 10.) Any one of these meaning shifts could be unobjectionable in some contexts: in most contexts it is unnecessary to make clear which of the alternative meanings is in use. Equally, these confusions are capable of generating fallacies. None of the books seriously explores the question of how to differentiate valid from unsound arguments in this connection, and we shall have to take it up later.

AMPHIBOLY

The word *amphiboly* means 'double arrangement': for many years it assumed an extra syllable and became 'amphibology' but this is just bad Greek, presumably short for the unpronounceable 'amphibolology'. Amphiboly is the same kind of thing as Equivocation except that the double-meaning occurs in a construction involving several words unambiguous in themselves. Copi (p. 76) cites the wartime austerity slogan

SAVE SOAP AND WASTE PAPER

and Thomas Gilby (*Barbara Celarent*, p. 254) was set on a train of amphibolous speculation by the sight, in Albermarle Street, of a door-plate announcing *The Society for Visiting Scientists*. The older examples of this Fallacy, some of which are still reproduced in textbooks, often involve fables about ambiguous prophecies, decrees, or inscriptions. To quote Abraham Fraunce again (f. 27):

> ... Amphiboly, when the sentence may be turned both the wayes, so that a man shall be uncertayne what waye to take, ... as that olde sophister the Devill deluded *Pyrrhus* by giving him such an intricate answere.
>
> > *Aio te, Aeacida, Romanos vincere posse.*
> > I now foretell the thing to thee
> >   which after shalbe knowne;
> > That thou, king Pyrrhus, once shalt see, the
> >   Romaines overthrowne.

Where this woord, overthrowne, may eyther bee the nominative case and appliable to king *Pyrrhus*: or the accusative, and attributed to the Romaynes.

The Latin can be translated either 'I say that you can conquer the Romans' or 'I say that the Romans can conquer you': *Aeacida* is another name for King Pyrrhus. The example is repeated by Joseph (p. 580), and by Fearnside and Holther (p. 162).[1] Copi (p. 75) and others give the apparently similar example of the Delphic Oracle's prophecy to Croesus, that if he went to war with the Persians, he 'would overthrow a mighty empire', fulfilled in the event by his overthrowing his own. This is not a clear case of Amphiboly, since it could be argued that Croesus was deceived as much by his own plain inability to see a possible implication. As Herodotus (*History*, Book I, ch. 91) reasonably says 'After an answer like that the wise thing would have been to send again to enquire which empire was meant.'

Schipper and Schuh (p. 53), following De Morgan (p. 287), take the numerical specification 'x equals two times four plus three' to be amphibolous, denoting either eleven or fourteen. Presumably the same applies, in the absence of brackets, or some special convention, to the symbolic formulation '$2 \times 4 + 3$'. Elaborate examples of mispunctuation have also been classified under this heading. The verse:

> I saw a comet drop down hail.
> I saw the clouds suck up a whale.
> I saw the sea within a glass.
> I saw some cider beat an ass.
> I saw a man full two miles high.
> I saw a mountain sob and cry.
> I saw a child with a thousand eyes.
> All this was seen without surprise.

can be converted into reasonable sense by repunctuation to put the stops after 'comet', 'clouds', 'sea', 'cider', 'man', 'mountain'

---

[1] It is given in the medieval *Viennese Fallacies* and is originally from Cicero, *De Divinatione*, Book 2, §116, where the prophecy was made not by the Devill but by Apollo. Cicero remarks that it is surprising that he should have spoken in Latin.

and 'child'.¹ A similar case of misplaced punctuation forms the central theme of the early Elizabethan play *Ralph Roister-Doister* in which a public letter-writer's punctuation of a dictated love-letter turns it into a stream of insults. This example is of some importance in literary history in that its citation by the logician Thomas Wilson was the clue that led to the ascription of the play, whose author had been unknown, to Nicholas Udall.²

But puns and anecdotes do not take the place of logical analysis. How many of the examples given are genuine cases of *fallacy*? The slogan 'Save Soap and Waste Paper' is not an argument at all, and even if some kind of invalid argument were erected on it there is very little likelihood that anyone would be persuaded of its validity. Again, Pyrrhus may or may not have *misconstrued* the prophecy made to him but, if he indulged in any kind of argument, what was mainly wrong with it was that, in believing what he thought he was told, he started from a false premiss. To get a good example of Amphiboly as it is defined by the textbooks we need to find a case in which someone was misled by an ambiguous verbal construction in such a way that, taking it to state a truth in one of its senses, he came also to take it to state a truth in its other sense. None of the examples so far quoted is of this character; and I regret to report that, in the books I have consulted, I have found no example that is any better.

COMPOSITION AND DIVISION

The Fallacy of Composition is described by Max Black (p. 232) as that in which 'what is true of a part is therefore asserted to be true of the whole'. His examples, however refer not to arguments from '*a* part' to 'the whole', but to arguments from '*all* the parts' to 'the whole'. Thus:

> Everybody in this city pays his debts. Therefore, you can be sure the city will pay its debts.

Schipper and Schuh explain (p. 50):

---

¹ I learnt this (I would not guarantee every word) from my grandfather. A slightly different version is given under the heading 'Fallacies of Punctuation', in Mercier, *A New Logic*, p. 374.
² Wilson, *The Rule of Reason* (1551); see Howell, *Logic and Rhetoric in England, 1500–1700*, pp. 30–1.

THE STANDARD TREATMENT   19

Names of collections or wholes are often used equivocally, in that such names, and their modifying adjectives, may refer to *each* of the members or parts of a class, or to the class as a *whole*. When an inference is made from properties of the parts of a whole, considered individually, to properties of the whole, considered organically or collectively, it is said that the fallacy of *composition* has been committed. For what is true of each of the parts may not hold true at all for the whole.

An example of this Fallacy occurs:

> ... when one reasons that each member of a football team is a good player, and that therefore the team as a whole is a good team.
> ... That a team consisting of all good players need not be a good team is obvious when we consider that all-star teams, composed presumably of the best individual college players, perhaps are seldom the best college teams in the nation.

Cohen and Nagel regard the Fallacy of Composition as one of a number of special forms resulting from the ambiguous use of words, and explain (p. 377):

> For the same word may have a different significance when applied to a totality than it has when applied to an element. Thus the fact that the soldiers of a given regiment are all 'strong' does not justify the conclusion that the regiment which they constitute is 'strong'. The word 'strong' does not mean the same in the two cases.

The Fallacy is sometimes mentioned in books on formal logic in a different connection; as when Quine, in a footnote to the distinction between 'class-membership' and 'class-inclusion' (*Mathematical Logic*, p. 189) says it occurred in a tentative form in traditional logic as

> ... a distinction between 'distributive' and 'collective predication', drawn to resolve the fallacies of composition and division (e.g. Peter is an Apostle, the Apostles are twelve, therefore Peter is twelve).

There are some differences to be noted between the above examples. Copi, in fact, says (p. 79 ff.) that there are two related varieties of fallacy with the same name. The first is 'reasoning fallaciously from the properties of parts of a whole to the properties of the whole itself', and

A particularly flagrant example would be to argue that since every part of a machine is light in weight, the machine 'as a whole' is light in weight.

The second is reasoning 'from properties possessed by individual members of a collection to properties possessed by the class or collection as a whole'. Quine's example comes under this rubric, and Copi distinguishes between the two kinds of predication involved, respectively, in the statements

Rodents have four feet.
and
Rodents are widely distributed over the earth.

The example of Cohen and Nagel about 'strong' soldiers making up a 'strong' regiment is quite different, however, if, as suggested, the word has two different senses at its two occurrences. The point of the distinction between distributive and collective predication is rather that, even without alteration of the sense of individual words, there may be two senses of a general sentence taken as a whole.

Oesterle (p. 255) gives the traditional example (Aristotle, *Soph. Ref.* 166a 33):

> Three and two are odd and even.
> Five is three and two.
> Therefore, five is odd and even.

Exactly how we are to analyse this argument, which may have been matter for furious debate between Socrates and some of his Sophist sparring partners but can hardly have aroused many strong emotions since the turn of the fourth century B.C., is not clear; but certainly, numerical addition, on which it depends, is not to be confused either with the putting together of parts into a whole or with the collection of individuals into a class. Actually, a careful reading of Aristotle suggests that he thinks of the Fallacy of Composition in a much more simple-minded way as arising in connection with different ways of grouping words together in a sentence.

In a paper in which he refers to Copi's account of the Fallacy, W. L. Rowe ('The Fallacy of Composition') raises the general question of the validity of arguments of the form 'All the parts of x have ϕ; therefore, x has ϕ', where ϕ is a property of some kind. In some cases, an argument of this form seems to be perfectly valid; thus 'All the parts of this machine are made of metal: therefore, this machine is made of metal', or 'All the parts of this chair are red: therefore, this chair is red'. In other cases, such as 'All the parts of this machine are light in weight: therefore this machine is light in weight', the invalidity is obvious. Rowe suggests that the distinction between the two cases rests in the fact that, when a fallacy is committed, there is an ambiguity in the property-word: that what it is, for example, for a machine as a whole to be light in weight is a different thing from what it is for a part of a machine to be light in weight. The word 'light', that is, is a relative word, and the fallacy that is committed is really a special case of the Fallacy of Equivocation. We could dispose of this suggestion by substituting the predicate 'weighs less than 1 lb' for 'is light in weight' but the problem remains of distinguishing the predicates for which the argument is valid from those for which it is not. Richard Cole[1] suggests that, so long as some predicates work and others do not, we must insist that, in strict logic, an additional premiss is necessary of the form (say)

> When all the parts of a chair are a certain colour, the chair is that colour.

and that it is only when the appropriate additional premiss is false that the argument fails. This seems too easy a move: used generally, it would make every fallacy of whatever kind a formal one. In the present case the issue will turn on whether the additional premiss is to be regarded as a necessary truth, and hence redundant, or not.

It is worth while remarking, however, that we sometimes need to distinguish *physical collections*, like piles of sand, from *functional collections*, like football teams, and these in turn from *conceptual collections*, like the totality of butterflies. Within each of these categories there are sub-categories. These distinctions are different from the ones already made.

The Fallacy of Division is supposed to be the reverse of the Fallacy of Composition, and to arise from the illicit replacement

---

[1] 'A Note on Informal Fallacies.' See also Bar-Hillel, 'More on the Fallacy of Composition.'

of a statement about a whole with a statement about its parts, rather than the other way round. Since to confuse A with B is the same thing as to confuse B with A it usually takes, at most, a little rewording to convert an example of Composition into an example of Division, or vice versa. Formally, an argument from A to B is always exactly as valid, or as invalid, as the argument from the contradictory of B to the contradictory of A; and, although non-formal features of an argument make it possible to distinguish the two, there is no particular point in doing so since the considerations involved are the same in the two cases. All the same distinctions – between part and whole, collective and distributive predication – are involved in the two cases. Nevertheless many books still list them separately and give separate examples. Copi has (p. 81):

> American Indians are disappearing.
> That man is an American Indian.
> Therefore that man is disappearing.

ACCENT

'Accents are divided into acute, grave and circumflex', wrote Peter of Spain (*Summulae Logicales*, 7.31), more gravely than acutely, under this heading in the thirteenth century. He was writing in Latin, which has never had written accents of this kind and, so far as anyone knows, has never been pronounced in such a way as to make them appropriate. In theory an acute accent (´) indicates a rising intonation, a grave (`) a falling one, and a circumflex (^ or ˆ) a rise and fall within the same syllable; and the Fallacy of Accent is supposed to arise from the confusion of words which are spelt alike but differ in spoken accentuation. Peter explains all this and then proceeds to cite Latin words and sentences which quite fail to exemplify it.[1] Altogether, treatments of the Fallacy of Accent, through the ages, provide an excellent example of adherence to superficial features of Aristotle's account coupled with complete neglect of its spirit. The history is worth tracing briefly.

[1] There was, however, some attempt to recognize accents in Latin. Aquinas, in his *Opusculum* on fallacies gives as examples: for acute, the middle syllable of *Martinus*; for grave, the middle syllable of *Dominus*; for circumflex, the word *Rom*.

Greek had no written accents when Aristotle wrote, but there is evidence that it was pronounced in such a way as to make such accents quite natural when they were introduced about the first century B.C.: the acute *did* indicate a rise in pitch, and so on. In fact, for all we know, the introduction of written accents could have been inspired by what Aristotle wrote, and have had the precise purpose of removing ambiguities such as can lead to the Fallacy. Aristotle (166a 39) says that an argument depending on accent is not easy to construct in unwritten discussion but easier in written discussions and poetry: the metre of Greek poetry depended on distinction between short and long vowels rather than intonation, and presumably declaimed poetry tended to be like written Greek in being accentless or, at least, a little indefinite in accent. The examples he gives, designed, perhaps, to be spoken in lectures, are poetic ones from Homer.

Medieval logicians, of whom Peter of Spain may be taken as typical, could not reproduce Aristotle's Fallacy in Latin but felt obliged to find a way of representing his account as meaningful. Peter gives two kinds of example. The first turns on a distinction between long and short vowels, as in the case of the word *populus* which, with short *o* means 'a people' and with long *o* 'a poplar'. Hence we have *Omnis populus est arbor: gens est populus: ergo gens est arbor* (Every poplar is a tree; a nation is a people; therefore a nation is a tree). This kind of ambiguity would be impossible in Latin poetry, which depends on the distinction between long and short vowels to establish its metre.

The second kind of example depends on reading two words as one or one as two. Thus the Latin word *invite* means 'against one's will' but *in vite* means 'in the grape-vine'; so that we have *Deus nihil fecit invite, ergo vinum non fecit in vite* (God made nothing against his will, therefore he did not make the wine in the grape-vine). This kind of example, however, is one that can occur in speech but cannot occur in writing, at least so long as wordbreaks are preserved. Either way it proves impossible to reproduce the features of Aristotle's Fallacy, which depends too closely on particular features of classical Greek to stand transplanting.

But what, now, of our modern writers? We would not expect misplaced reverence for Aristotle or medieval tradition to be a general feature of modern writing. Logicians, however, have been unwilling to part with the Fallacy of Accent, and it is still in the lists in at least half of the books I have studied.

In English we do not rely much on rise or fall of intonation and we have few words which can be pronounced alternatively with long or short vowel-sounds. There remains stress: we do distinguish, in speech but not normally in writing, between words differently stressed, as in the case of 'in*cense*' and '*in*cense'. (This is Oesterle's example, p. 255. To *in*cense a person is, presumably, to surround him with perfumed smoke. This may or may not in*cense* him.) It is probably rather rarely, in practice, that anyone ever perpetrates or is deceived by a fallacious argument that turns on this confusion, but at least it ought to be possible to produce textbook-style examples which have approximately the features of Aristotle's originals. Unfortunately, stress also has another function in English or, for that matter, any other language: it connotes emphasis. Logicians are evidently incapable of making distinctions of this sort, and go overboard for examples in which changes of emphasis on the various *words* in a *sentence* alter the meaning of the sentence as a whole. Copi (p. 76) says

> Consider the different meanings that are given according to which of the italicized words is stressed in the injunction:
>
> *We should* not *speak* ill of *our friends*.
>
> When read without any undue stresses, the injunction is perfectly sound. If the conclusion is drawn from it, however, that we should feel free to speak ill of someone who is *not* our friend, then this conclusion follows *only* if the premiss has the meaning it acquires when its last word is accented. But when its last word is accented, it is no longer acceptable as a moral law, it has a different meaning, and is in fact a different premiss. The argument is a case of the fallacy of *accent*.

I am sorry to keep picking on Copi. It is perhaps because his account is better than most that one is inclined to regard him as a spokesman. Similar examples are given by Oesterle, by Schipper and Schuh (p. 52) and others. The original of this *genre* is in De Morgan (p. 289).

But we have not yet finished. We are on a slippery slide, and now that verbal emphasis has been allowed as relevant, *any* kind of emphasis may be called in. Copi cites the false emphasis that may be given typographically by newspaper headlines or in advertisements. Schipper and Schuh say

> The fallacy of *special pleading* or *half-truth* may be considered a distinctive kind of illegitimate accent. For if one emphasizes only those circumstances favourable to his own case, and conveniently forgets the unfavourable circumstances, he is wrongfully accenting or stressing only part of the truth. It must be admitted that special pleading is the stock in trade of the legal profession. One wonders indeed how an attorney, especially one who pleads his cases in court, could possibly build a successful practice without persistently and cleverly resorting to this fallacy.

We have come a long way since Aristotle.

The phenomena referred to by Copi, Schipper and Schuh, and others are, no doubt, worth mentioning as important elements in a theory of argumentation. They are, however, very various in character and are not to be summed up in just an example or two; least of all, with the misleading title 'fallacy of accent'. As Gilby says (p. 255):

> We touch here the persuasions of rhetoric, for there are accents other than those of sound: subtler psychological stresses can be set up by the position, or repetition, or emotional tone of words, by the archness in the pauses of a radio news announcer and the overcharging of a thesis by a research scholar. The mind is an opal, rather than a clear crystal; its words are echoing chords, not single pings.

The best thing the modern logician might do would be to give this field some explicit, separate attention: some of them do, though a satisfactory account is hard to find. The next best is to omit all mention: it is only fair to say that many of them do this.

FIGURE OF SPEECH

This is the last of Aristotle's six Fallacies Dependent on Language. It consists in being deceived by the misleading structure or etymology of a word. Very few modern writers even bother to mention it. Their difficulty, if they do so, is to find serious examples of it. Joseph (p. 584) illustrates it with the apparent passivity of 'I am resolved what to do', which suggest that a man's resolution is not his own free act, and with the apparent negativity of 'important', which has the same prefix as the genuinely negative 'imperturbable' and 'impenitent'; and he refers, in a footnote, to a lady who once observed: 'The question is, is he a postor or an impostor?' Fearnside and Holther (p. 168), without referring to Figure of Speech explicitly, have a section entitled 'Misuses of Etymology' in which some similar arguments are explored. We do not need to examine these examples very closely.

It was given to J. S. Mill to make the greatest of modern contributions to this Fallacy by perpetrating a serious example of it himself. This was what the textbook writers were waiting for, and he is widely quoted. He said (*Utilitarianism*, ch. 4, p. 32)

> The only proof capable of being given that an object is visible, is that people actually see it. The only proof that a sound is audible, is that people hear it: and so of the other sources of our experience. In like manner, I apprehend, the sole evidence it is possible to produce that anything is desirable, is that people do actually desire it.

But to say that something is visible or audible, is to say that people *can* see or hear it, whereas to say that something is desirable is to say that it is *worthy of* desire or, plainly, a good thing. Mill is misled by the termination '-able'.

How did the phrase 'figure of speech' get into the language? Aristotle was apparently not the first to use it, but borrowed it from earlier rhetorical teachings. It was probably first used by the sophist Gorgias.[1] It has had a variety of meanings in Rhetoric and Grammar before giving us, in modern times, the word 'figurative' to describe metaphor. Logicians usually do not know what to say about metaphor and are content to let grammarians and ordinary men have this use to themselves.

ACCIDENT

Leaving Aristotle's first group of Fallacies we turn to Fallacies Outside Language. On starting to look at examples we could be

---

[1] See Diodorus Siculus, *History*, XII, 53, 4. Cf. Aristotle, *Rhetoric* 1408b 20, 1410b 28, and *Poetics* 1456b 9.

pardoned for thinking that the change of genus is not going to make much difference. A much-quoted example which is as old as Plato's *Euthydemus* is

> This dog is yours
> This dog is father
> Therefore, this dog is your father.

This is a verbal pun as flagrant as any of the Fallacies Dependent on Language. Cohen and Nagel, in fact, put the Fallacy of Accident and the (following) Fallacy of *Secundum Quid* under the heading 'Semilogical or Verbal Fallacies'. This would have been inappropriate for Accident as originally understood, though somewhat less so for *Secundum Quid*. It is not at all out of line with some modern treatments.

The name 'Accident' is a doctrine-bound one which turns on a particular analysis by Aristotle of a class of controversial examples. In theory these fallacies arise from taking an accidental property to be an essential one, and it is this that most of the books take as their starting-point. Unfortunately it is often difficult to say of a property whether it is 'essential' or not, and few people these days would be prepared to go so far as to maintain generally that the essential properties of every kind of thing can be uniquely specified. Yet all the most common examples depend on this assumption. Oesterle, for example, argues that racialism in politics is due to the mistaken belief that skin colour and similar racial characteristics are 'essential' properties of those who bear them. It is difficult to know what this means: if it is that these characteristics should never be taken into account for any purpose, it is plainly at variance with common sense and would rule out good differential treatment along with bad. Moreover, two can play at the game of Essentialism and it is not at all clear how someone who claims that a given property is not an essential one will be able to argue against someone else who claims that it is. A charge of fallacy may simply be returned to the sender.

Speaking relatively instead of absolutely, we might say that racial differences are essential for certain purposes, unessential for others. But if we speak this way, we can no longer provide so simple a rationale for the Fallacy of Accident.

Consequently an alternative, slightly different rationale is often provided. Copi says (p. 63).

> The fallacy of *accident* consists in applying a general rule to a particular case whose 'accidental' circumstances render the rule inapplicable.

Should we pay our debts? Yes, as a general rule, but there are exceptional situations in which the obligation lapses. The word 'accidental', however, is not really appropriate as a characterization of the exceptional situations, and it could be omitted without loss from Copi's description, as he nearly acknowledges by placing it between quotes. What we call a *general* rule, if it is not (as we might prefer) an absolutely universal one, is one which holds most of the time, in general, normal or usual circumstances, but this is not to say that it depends on 'essential' properties of things, or that exceptions are 'accidents'.

This change in interpretation brings the Fallacy of Accident close in character to the next one in the list. We shall return, in the next chapter, to Plato's paternal dog.

'SECUNDUM QUID'

Secundum Quid, Greek παρὰ τὸ πῇ, means 'in a certain respect' and refers to the qualifications which may be attached to a term or generalization. Fallacies *secundum quid* are those which involve neglect of necessary qualifications. Copi's quoted description of the Fallacy of Accident puts it firmly in this category. It is sometimes said, following De Morgan, that the two kinds of fallacy are converses and that, whereas Accident is argument from general to inappropriate particular, *Secundum Quid* is invalid argument from particular to general. However, in view of the ease with which an argument may be cast into different forms, the distinction is a little artificial. If someone argues that alcohol is bad because it causes drunkenness there may be, in his argument, some kind of movement of meaning; but the direction of the movement is not clear until we determine which of the two statements 'Alcohol is bad' and 'Alcohol causes drunkenness' is to be regarded as a strictly universal generalization and which as a qualified one.

Underlying the difficulty is a more serious one, which we shall have to face in due course. 'Hasty generalization', as Copi (p. 64) describes what is apparently intended to be this Fallacy, is a common logical sin, but logicians themselves are far from being agreed on criteria for what are called 'inductive' arguments. It has been seriously argued by Hume and others that *every* argument from particular cases to a general rule must be fallacious, since it is impossible to survey, and to know that one has surveyed, all the particular cases the generalization covers. We seem to have a double logical standard: one set of criteria for deductive arguments and another for inductive ones. If so, there are two sets of criteria for fallacy as well'. Most or all generalizations are *deductively* unjustified; but we have a need for other criteria which will enable us to distinguish between the too-hasty ones and the relatively reliable. As long ago as the *Port Royal Logic*[1] an attempt was made to supplement the traditional account of fallacies with some discussion of unjustified inductions, and Mill's account of fallacies (*System of Logic*, Book 5) is strongly oriented, at least superficially, in this direction. For whatever reason, these attempts to remedy a deficiency of the traditional scheme have been dropped by most recent writers, and all that remains is some brief reference under other headings.

The Fallacy of *Secundum Quid* was never designed to carry this kind of load and does so uneasily. The distinction between a qualified and a strictly universal generalization is of minor importance in discussions of induction and vice versa. We shall need, later, to investigate the relation between them; but only after putting both into historical context.

An interesting, and entirely typical, illustration of the ossification of the traditional treatment of fallacies in modern times concerns the example:

> What you bought yesterday, you eat today
> You bought raw meat yesterday
> Therefore, you eat raw meat today.

which appears first in the twelfth-century *Munich Dialectica*,[2]

---

[1] Arnauld, *L'Art de Penser* (1662); known as the *Port Royal Logic*.
[2] De Rijk, *Logica Modernorum*, vol. 2, part 2, p. 580. For reasons to be discussed later, this example was classified in the Middle Ages as Figure of Speech.

and is usually regarded as an example of *Secundum Quid* but is given in some modern books (Cohen and Nagel, p. 377; Copi, p. 63) as an example of Accident. We have already noticed that there is not much difference between the two categories. However, the reclassification of the example can almost certainly be put down not to conscious design but to a historical mistake. Writers of textbooks take their examples one from another. De Morgan had written (pp. 291–2):

> 1. The *fallacia accidentis*; and 2. That *à dicto secundum quid ad dictum simpliciter*. The first of these ought to be called that of *à dicto simpliciter ad dictum secundum quid*, for the two are correlative in the manner described in the two phrases. The first consists in inferring of the subject with an accident that which was premised of the subject only: the second in inferring of the subject only that which was premised of the subject with an accident. The first example of the second must needs be 'What you bought yesterday you eat today . . . [etc.]' . . . Of the first, we may give the instance 'Wine is pernicious; therefore, it ought to be forbidden'.

It is clear both that De Morgan takes the 'raw meat' example to be a case of *Secundum Quid* proper, and that, on a careless reading, he *could* have been understood as taking it to be a case of Accident. When it is added that the mistake is not found before De Morgan and is often found after him, the possibility that this is its origin becomes a strong presumption.

I cannot leave the Fallacies of Accident and *Secundum Quid* without a comment on the way in which people invoke them in order to seek a logical sanction for their personal prejudices. These two heads of fallacy seem ideally suited to bolster any preconceived notion anyone may happen to have. We see something of this, even though we may agree with the conclusions, in Oesterle's analysis of racialism. A similar example, from the same author (p. 257), is given for *Secundum Quid*:

> In general, this fallacy consists of using a proposition, which has a qualified meaning, as though it applied in all circumstances and without restriction. One thus argues fallaciously that the commandment 'Thou shalt not kill' forbids fighting for one's country. But the meaning and context of the commandment forbids killing an innocent person unjustly, that is, murdering.

Does it? This is much too easy a way out. Let us admit, if we must, that the Ten Commandments are not to be taken literally; but, if someone wants to pay lip-service to a principle while making convenient exceptions, at least he should not be allowed to enlist the authority of Logic.

'IGNORATIO ELENCHI', MISCONCEPTION OF REFUTATION

The traditional term *Ignoratio Elenchi* means 'ignorance of refutation'. Oesterle translates it 'ignoring the issue' and Black, Copi, and Schipper and Schuh as 'irrelevant conclusion'. Aristotle (167a 21) shows that he means it to refer to cases in which, through lack of logical acumen, an arguer thinks he has proved one thing but has at best proved something else. 'The journey has been safely performed', says Sidgwick (p. 182), 'only we have got into the wrong train.' So described, this category can be stretched to cover virtually every kind of fallacy; or it can be restricted to clear cases of misintepretation of the thesis. Schipper and Schuh treat it frankly as a rag-bag and say (p. 36):

> Many arguments in which the premises are irrelevant to the conclusion cannot be classified properly under any of the foregoing headings. It will prove convenient, therefore, simply to call these fallacies of *irrelevant conclusion*, using this as the title of a miscellaneous or catchall category of fallacies of relevance.

Schipper and Schuh's other 'fallacies of relevance' are the *ad hominem*, the *ad verecundiam* and other arguments *ad* something-or-other. We shall deal with these separately. They are peculiar in that they represent specious attempts, usually emotional, to persuade an audience, and Aristotle's examples are not like them at all. Copi says (p. 69):

> An argument may be stated in cold, aseptic, neutral language and still commit the fallacy of irrelevant conclusion.

However, I have not been able to find in modern books any examples fitting this description. The prosecuting counsel in a murder case argues irrelevantly that murder is a horrible crime (Copi); a liberal education is not practical because it does not result immediately in a cash dividend (Oesterle); opponents of federal health insurance plans demonstrate the dangers (!) of the socialized medicine programmes of Great Britain or Sweden (Schipper and Schuh). Whether the Lockean categories are included or not, the name *Ignoratio Elenchi* inadequately characterizes these examples; and since there are no other modern examples for it to characterize it has no modern justification.

I shall later prefer the term 'Misconceptions of Refutation' in referring to the earliest accounts of this Fallacy.

BEGGING THE QUESTION

The origin and meaning of the phrase 'beg the question', and its Latin and Greek counterparts, have proved a problem to many writers and their readers. Some interesting etymologies have been proposed. I recall that I used to think 'beg the question' was a corruption of 'beggar the question'; not unlike a twelfth-century colleague who had trouble making sense of the Latin *petitio* and transcribed it *repetitio*.[1] An eighteenth century French textbook[2] refers to

...*pétition de principe* from the Greek word πέτομαι, which means to *fly towards something,* and the Latin word *principium*, which means *beginning;* thus to commit a *pétition de principe* is to run back in new words to the same thing as was originally in dispute.

In fact 'beg the question' is a reasonably accurate translation of Aristotle's original Greek τὸ ἐν ἀρχῇ αἰτεῖσθαι (– in place of αἰτεῖσθαι Aristotle also sometimes uses the word λαμβάνειν 'to assume' –) provided we suitably interpret the word 'question': the phrase in the original actually means something like 'beg for that which is in the question-at-issue'. The Latin *principium petere* is the vulgate translation of this, and 'beg the question' has been the accepted English one at least since the sixteenth century.[3] Nevertheless Webster's Dictionary still translates *petitio*

[1] De Rijk, *Logica Modernorum,* vol. I, p. 100.
[2] Du Marsais, *Logique* (p. 81); see Mansel's edition of Aldrich, *Artis Logicae Rudimenta,* p. 201.
[3] I owe some enlightenment on this point to A. C. W. Sparkes, who finds the phrase 'demanding the thing in question' used in 1577. See his note 'Begging the Question'. Abraham Fraunce has 'the requesting of the thing in controversie' in 1588.

principii as 'postulation of the beginning' and, to balance one sin with another, gives 'evade' as a synonym of 'beg'.

The arrangement of fallacies in Schipper and Schuh (pp. 55–60) suggests that these authors fail to realize that 'beg the question' is connected with *petitio principii*. Under the general heading 'Fallacies of Presumption' they explain:

> These presumptive arguments are sometimes called *question-begging*, because by smuggling the conclusion into the wording of the premises, they beg or avoid the question at issue in the argument.

(Note the gloss 'or avoid'.) There are four separate kinds of 'Fallacy of Presumption' listed, including 'Complex Question', yet to be discussed. The fourth is 'Circular Reasoning',

> ... sometimes called the 'vicious circle argument' or by its Latin name, *petitio principii*...

This is an involved confusion of terminology. Apart from anything else, a *vicious* circle is one involving self-contradiction or self-defeat, as in the paradox of the Liar.

Why 'beg'? We shall understand this better when we come to put it in the context of disputation on the Greek pattern, as Aristotle originally intended it. If one person sets out to argue a case to another he may *ask to be granted* certain premises on which to build. The Fallacy consist in asking to be granted the question-at-issue, which one has set out to prove. What makes the phrase confusing is that modern examples are so seldom presented as specimens of lifelike disputation. Keynes[1] says, making a distinction that has not occurred before in this discussion, that Begging the Question is a *fallacy of proof* rather than a *fallacy of inference:* apparently to ward off the criticism that there is nothing wrong with inferring something from itself unless it is accompanied by a claim to have proved something. If there is really a distinction between fallacies of proof and fallacies of inference it should, of course, be made generally and used as a feature of classification, but nobody has done this.

---

[1] Keynes, *Studies and Exercises*, p. 425. The distinction had been made in the thirteenth century: Peter of Spain wrote (*Summulae* 7.54) 'It is to be noted that this fallacy is no impediment to an inferring syllogism, but only to a probative one.'

The name 'Begging the Question' is often extended to cases in which, although the precise point in dispute is not itself assumed as a premiss, something equally questionable is assumed in its place. Abraham Fraunce has (f. 28)

> *Petitio principii*, then, is eyther when the same thing is prooved by it selfe, as, *The soule is immortall, because it never dyeth*: Or when a doubtful thing is confirmed by that which is as doubtfull, as
> *The earth mooveth*
> *Because the heaven standeth still.*

Whately (*Elements of Logic*, Bk. II, § 13) is a little more explicit when he defines it as that

> in which one of the Premises either is manifestly the same in sense with the Conclusion, or is actually proved from it, or is such as the persons you are addressing are not likely to know, or to admit, except as an inference from the Conclusion ...

Curiously, the commonest example in the modern books has a twist to it which prevents either of these definitions from clearly capturing it. Cohen and Nagel (p. 379) say

> ... it would be arguing in a circle to try to prove the infallibility of the Koran by the proposition that it was composed by God's prophet (Mahomet), if the truth about Mahomet's being God's prophet depends upon the authority of the Koran.

Black (p. 236) has, in essence, the same example in the shape of a dialogue between a man and his bank manager: 'My friend Jones will vouch for me.' 'How do we know *he* can be trusted?' 'Oh, I assure you he can.' Copi refers to the argument that Shakespeare is greater than Spillane because people with good taste prefer him, the good taste being demonstrated by ... etc., etc.; and Oesterle (pp. 257–8) has a partly similar example concerning proving the existence of God from the idea of God as existing in the human mind and arguing the reliability of human powers of knowing from the existence of God. All these arguments are at least partly *arguments from authority*; and, elsewhere in their discussions, most of the authors concerned point out that arguments from authority are to be distinguished from arguments on the merits of a case and are even, perhaps, essentially fallacious themselves. We shall discuss the merits of arguments from authority later; but this is a complication that one would expect to be absent from textbook prototypes of another fallacy.

However, by far the most important controversy surrounding *petitio principii* concerns J. S. Mill's claim that *all* valid reasoning commits this Fallacy. Cohen and Nagel touch on this when they say (p. 379):

> There is a sense in which all science is circular, for all proof rests upon assumptions which are not derived from others but are justified by the set of consequences which are deduced from them.... But there is a difference between a circle consisting of a small number of propositions, from which we can escape by denying them all or setting up their contradictories, and the circle of theoretical science and human observation, which is so wide that we cannot set up any alternative to it.

Apparently a fallacy is not objectionable so long as it is big enough. This is, however, a serious philosophical assertion and we must discuss it as such in a later chapter.

AFFIRMING THE CONSEQUENT

This Fallacy, as Aristotle explains (167 b1),

> arises because people suppose that the relation of consequence is convertible. For whenever, suppose A is, B necessarily is, they then suppose that if B is, A necessarily is.

A relation is *convertible* if its two terms can be validly interchanged. We need not hang too much on Aristotle's word 'necessarily'. The ordinary form of reasoning from *S implies T and S is true* to *T is true* is commonly called *modus ponens*; and the Fallacy of the Consequent is generally regarded as a backwards version of this, from *S implies T and T is true* to *S is true*. Copi, for example, says (p. 225):

> One must not confuse the valid form *modus ponens* with the clearly invalid form displayed by the following argument.
>
> > If Bacon wrote *Hamlet*, then Bacon was a great writer.
> > Bacon was a great writer.
> > Therefore Bacon wrote *Hamlet*.

This argument differs from *modus ponens* in that its categorical premiss affirms the consequent, rather than the antecedent, of the hypothetical premiss. Any argument of this form is said to commit the *Fallacy of Affirming the Consequent*.

The phrase 'affirming the consequent' is not from Aristotle and goes back, I think, no further than J. N. Keynes who, though he does not use it as a tag, says (*Studies and Exercises*, p. 353) 'It is a fallacy to regard the affirmation of the consequent as justifying the affirmation of the antecedent.' From the same source we get the concept of the Fallacy of Denying the Antecedent. Copi's example would commit this Fallacy if it were turned round to read

If Bacon wrote *Hamlet* then Bacon was a great writer.
Bacon did not write *Hamlet*.
Therefore Bacon was not a great writer.

In the previous case the consequent 'Bacon was a great writer' of the hypothetical premiss was affirmed as a second premiss and the antecedent 'Bacon wrote *Hamlet*' invalidly inferred from it: in this case the antecedent is denied in the second premiss and the denial of the consequent is equally invalidly inferred. Examples very like these were given in antiquity by the Stoics, as reported by Sextus Empiricus and others.[1]

All of our modern books identify and name this Fallacy but only one, the traditionalist Oesterle, lists it in order with the other Fallacies. The others treat it along with inferences of the calculus of propositions. The divorce between Fallacies and the rest of Logic could hardly be more complete. As soon as a Fallacy has some relation to the rest of Logic it is removed from its place in the chapter on Fallacies!

There is a subtle difference between Aristotle's treatment and that of the Stoic and modern writers, in that Aristotle gives us not examples with a hypothetical 'If . . . then . . .' formulation but rather examples consistent with his 'syllogistic' logic of class-terms. Even so we might reasonably ask why he fails to treat Consequent as a 'formal' fallacy in the way the modern books do; and we shall investigate this question in due course. The modern

---

[1] See Sextus Empiricus, *Outlines of Pyrrhonism*, II, §147; *Against the Logicians*, II, § 420.

books are, of course, quite consistent in treating it separately from the others, since it is sufficiently proscribed by the rules they all give for propositional inferences. What is less clear is why it is still singled out at all. Every invalid inference-schema of the propositional calculus – or, for that matter, of other logical systems – could, in theory, be dignified with a special name and treated similarly, yet we do not hear of any others. Why not the Fallacy of Inferring the Conjunction of Two Propositions from their Material Equivalence; or, say, the Fallacy of Distributing Quantifiers without regard to Negation Signs?

FALSE CAUSE

Copi, after explaining that various analyses have been given to this Fallacy, says (p. 64)

> We shall regard any argument that incorrectly attempts to establish a causal connection as an instance of the fallacy of *false cause*.

We are now classifying fallacies according to the kind of conclusion they have! Schipper and Schuh continue (p. 35)

> In practice, however, the false-cause fallacy has come to mean a specific kind of illicit argument, that is, one which involves an inference from a merely temporal sequence of events to a causal sequence. The Latin expression for this fallacy precisely describes its nature: *post hoc, ergo propter hoc* – after this, therefore because of this.

Even so, this is puzzling. If we know that B *always* occurs after A we are well on the way to setting up a causal law, and the precise difference between necessary connection and constant conjunction has been a matter for debate among philosophers at least since Hume. Hasty or unwarranted generalizations have, moreover, already been proscribed in this list in the name of the Fallacy of *Secundum Quid:* yet the examples given by the quoted authors – herb medicines 'curing' colds, excursions under ladders ending in broken legs, rabbit's foot tokens securing pay rises – are clear examples of hasty generalizations whose main fault is that they proceed from temporal conjunctions which are actually found *not* to be constant. As in the case of some previous

Fallacies, the mystery is why logicians should think it worth while to preserve an incoherent tradition.

The interpretation of Copi and Schipper and Schuh (and Oesterle) has some sanction from Aristotle in that a passage in his *Rhetoric* gives it (1401b 30); and the tradition that backs it is a long one. We shall see later, however, that Aristotle's main account of this Fallacy is quite different and makes more sense, referring to a fault which can occur in arguments of the kind known as *reductio ad impossibile*.

MANY QUESTIONS

The Fallacy of Many Questions or, more comprehensibly, Fallacy of the Complex Question is most commonly illustrated by the question 'Have you stopped beating your wife?', which seems designed to force ordinary non-wife-beaters into admission of guilt. Another example is Charles II's (probably apocryphal) question to the Royal Society. As reported by Joseph (p. 597) this asked

> Why a live fish placed in a bowl already full of water did not cause it to overflow, whereas a dead fish did so; . . .

There are various different versions of the story, and Joseph's is not the original one. Whateley (Book 3, § 14) has

> . . . the Royal Society were imposed on by being asked to *account for* the fact that a vessel of water received no addition to its weight by a dead fish put into it; . . .

What is probably the earliest extant version is given by Isaac D'Israeli.[1] On the occasion of the creation of the Royal Society the King dined with its members and, towards the close of the evening, admitted,

> with that peculiar gravity of countenance he usually wore on such occasions, that among such learned men he now hoped for a solution to a question which had long perplexed him. The case he thus stated: 'Suppose two pails of water were fixed in two different scales that were equally poised, and which weighed equally alike,

---

[1] *Quarrels of Authors* (1814), chapter on 'The Royal Society', p. 341.

and that two live bream, or small fish, were put into either of these pails, he wanted to know the reason why that pail, with such addition, should not weigh more than the other pail which stood against it.' Every one was ready to set at quiet the royal curiosity; but it appeared that every one was giving a different opinion. One, at length, offered so ridiculous a solution, that another of the members could not refrain from a loud laugh; when the King, turning to him, insisted, that he should give his sentiments as well as the rest. This he did without hesitation, and told His Majesty, in plain terms, that he denied the fact! On which the King, in high mirth, exclaimed 'Odds fish, brother, you are in the right!'

Nevertheless, in this version, the problem is a not entirely trivial one in fluid dynamics. Max Black has suggested for comparison[1] the case of the weight of a cage with covered floor and sides and containing a bird in flight.

However this may be, Charles's question, like 'Have you stopped beating your wife?', carries with it a presumption which may prejudice an attempt to give a straightforward answer. A presumption is unimportant only when it is clearly true; and when it is not true the question is properly answered only with an objection which denies or attacks the presumption, in some such form as 'I have never beaten my wife', or 'Before offering an explanation, let us be sure there is a fact to be explained.' This having been said, it remains to ask what relevance these examples have in a list of fallacies. A fallacy, we must repeat, is an *invalid argument*; and a man who asks a misleading question can hardly be said to have argued, validly or invalidly, for anything at all. Where are his premisses and what is his conclusion?

We shall find a resolution of this difficulty, as with some previous ones, when we come to consider ancient Greek patterns of public debate. Nevertheless it is a long time since there have been any ancient Greek public debates, and logicians have not been quick to adjust to their discontinuance. Let us just note one even less suitable example. Joseph (p. 598), copied by Copi (p. 67), refers to

> the custom of 'tacking' in the American legislature. The President of the United States can veto bills, and does veto them freely; but he can only veto a bill as a whole. It is therefore not uncommon for

[1] In a talk in Sydney in 1966.

the legislature to tack on to a bill which the President feels bound to let pass a clause containing a measure to which it is known that he objects; so that if he assents, he allows what he disapproves of, and if he dissents, he disallows what he approves.

But Congress, in so doing, could not, by any stretch of imagination, be regarded as guilty of a *fallacious argument*. It is simply a quirk of constitutional law that the President is forbidden from sorting things out by 'dividing the question', as is accepted practice in debates within Congress itself.

Oesterle, discussing the wife-beating example, says (p. 259):

> This type of question is called a 'leading question' and is ruled out of court in legal debates.

This is another confusion. At law, *any* question which is so definite as to call merely for the selection of one of a short list of predetermined possible answers is called a 'leading question'; but leading questions are generally admitted without restriction in cross-examination of a witness by an opposing counsel and are prohibited only during the direct examination of a witness by the counsel responsible for calling him.

Many Questions is the last of the Fallacies in Aristotle's list. Several of the later items in this list have led us to question the coherency of the classification. In the case of some of them, including this one, the word 'fallacy' seems to be misdescription. I shall later argue that this is the case and that many of the discussions in modern logic books do a different job from the one they purport to do. It is of some interest that the phenomenon of the complex question also receives a mention in connection with recent work on the formal logic of questions, where it is essential to recognize that questions may – and, in fact, usually do – involve presumptions and that there are various differently appropriate kinds of answer in such cases. Work of this kind is a contribution to the theory of the use of language in practical situations: what Carnap calls Pragmatics and what we shall find reason to call Dialectic. It may be in this field that the discussions surrounding some of these so-called Fallacies find their true modern home.

'AD HOMINEM'

As we turn from Aristotle's list to later additions our attention is claimed first by a group of alleged Fallacies known as the *argumentum ad hominem*, the *argumentum ad verecundiam*, the *argumentum ad misericordiam* and the *argumenta ad ignorantiam, populum, baculum, passiones, superstitionem, imaginationem, invidiam* (envy), *crumenam* (purse), *quietem* (repose, conservatism), *metum* (fear), *fidem* (faith), *socordiam* (weak-mindedness), *superbiam* (pride), *odium* (hatred), *amicitiam* (friendship), *ludicrum* (dramatics), *captandum vulgus* (playing to the gallery), *fulmen* (thunderbolt), *vertiginem* (dizziness) and *a carcere* (from prison). We feel like adding: *ad nauseam* – but even this has been suggested before.[1] Most of our books mention a few of them, though I do not know of any that treats them all. Fearnside and Holther say (p. 94)

> These Latin terms show how long these fallacies have been recognised; a naïve person might be surprised we still have them with us.

These authors may be interested to know that the *genre* was invented by Locke and that all but a few of the names are from the nineteenth and twentieth centuries. Incidentally, as we shall see later, Locke does not clearly say that he regards them as Fallacies.

According to modern tradition an argument *ad hominem* is committed when a case is argued not on its merits but by analysing (usually unfavourably) the motives or background of its supporters or opponents. For example, Cohen and Nagel say (p. 380)

> ... attempts have been made to refute some of Spinoza's arguments as to the nature of substance, or as to the relation of individual modes to that substance, on the ground that they were advanced by a man who had separated himself from his people, a man who lived alone, was intellectualist in temper, and so on.

Or one might, says Joseph, condemn Home Rule for Ireland on the grounds that Parnell was an adulterer. As already mentioned, writers wedded to Aristotle's classification often fit this Fallacy in under *Ignoratio Elenchi*; and since almost any fallacy at all might be put under this heading we can have no objection. The

[1] By F. H. Bradley in *Appearance and Reality*, p. 35.

main question, however, is not one of classification but of whether arguments *ad hominem* are genuinely fallacious. Several of our authors express doubts. Joseph says (p. 591)

> A barrister who meets the testimony of a hostile witness by proving that the witness is a notorious thief, though he does less well than if he could disprove his evidence directly, may reasonably be considered to have shaken it; for a man's character bears on his credibility. And sometimes we may be content to prove against those who attack us, not that our conduct is right, but that it accords with the principles which they profess or act upon.

Copi distinguishes two varieties of *ad hominem* argument which he calls 'circumstantial' and 'abusive'. Circumstantial arguments are not always invalid, though it is not clear when they are and why. Purely abusive arguments, on the other hand, are not arguments at all, though Copi does not say so. The further problem arises of distinguishing pure abuse from relevant circumstantial comment.

Fearnside and Holther (p. 94), perhaps following Whately (see our p. 174 below), contrast *ad hominem* with *ad rem*, which means 'to the point' or 'relevantly'. The contrast is sound enough in some contexts, but the latter is a legal term and has no historical connection with the former.

'AD VERECUNDIAM'

Verecundia means 'shame' or 'shyness' or 'modesty' but an argument *ad verecundiam* is usually, not quite appropriately, regarded as an argument which rests on respect for authority; what Bentham calls the 'wisdom of our ancestors, or Chinese argument'. Presumably I can respect authorities without being ashamed, shy or (particularly) modest. Copi says (p. 62)

> Advertising 'testimonials' are frequent instances of this fallacy. We are urged to smoke this or that brand of cigarettes because a champion swimmer or midget auto racer affirms their superiority.

Once again (as Copi himself insists) we find a species of argument that is not clearly fallacious. An argument of the form

X is an authority on facts of type T
X said S, which is of type T
Therefore, S is true.

may leave something to be desired where deductive validity is concerned but the premisses, if true, do at least lend the conclusion support. The trouble with the quoted examples is that what corresponds with the first premiss is false. At various historical periods arguments from authority have been especially disliked, but this has been more because some particular 'authorities' have been distrusted than because there is anything wrong with an argument proceeding from a premiss which truly asserts expertise. Historically speaking, argument from authority has been mentioned in lists of valid argument-forms as often as in lists of Fallacies.[1]

'AD MISERICORDIAM'

Copi (p. 58) quotes the speech of barrister Clarence Darrow in the defence of a union member charged with criminal conspiracy. *Misericordia* means 'pity', and this appeal to pity 'was sufficiently moving to make the average juror want to throw questions of evidence and of law out the window'. The fallacious argument proceeds by engaging the hearer's emotions to the detriment of his good judgement.

We readily recognize this syndrome and it seems carping to object. However, more depends on a lawsuit, or a political speech, than assent to a proposition. A proposition is presented primarily as a guide to action and, where action is concerned, it is not so clear that pity and other emotions are irrelevant.

'AD IGNORANTIAM'

'The *argumentum ad ignorantiam* is illustrated by the argument that there must be ghosts because no one has ever been able to prove that there aren't any.' However, 'this mode of argument is not fallacious in a court of law, because there the guiding principle is that a person is presumed innocent until proven guilty' (Copi, p. 57). We also have guiding principles in our everyday affairs; but it must be a strange form of argument that is now valid, now

---

[1] Starting with Aristotle in *Rhetoric* 1398b 18.

invalid, according as presumptions change with context. Actually, belief in ghosts in the absence of evidence itself offends against a commonly revered philosophical principle, the Ockhamist 'Entities are not to be multiplied unnecessarily'; and perhaps that is what is really wrong with it.

The *argumentum ad ignorantiam* is nominally an appeal 'to ignorance'; but it is not quite clear, from some of the examples given, that it does not consist alternatively of a browbeating of ignorant people into accepting the views of the speaker.

BACULUM, POPULUM, ODIUM, ETC.

The other 'arguments *ad*' are more rarely mentioned. Most of them are appeals to one or other specified emotion. The *argumentum ad populum* is an appeal to popular favour, which, to preserve uniformity, must be purely emotional, though it is not clear from its name that it does not consist of the purest valid reasoning, and only an anti-democrat could unhesitatingly assume the contrary.

The *argumentum ad baculum* is dignified with a paragraph by Copi (p. 53): *baculum* means 'stick', so this is argument by threat. The preposition '*ad*' clearly means many different things. We need not, however, pursue these argument forms further.

FORMALLY INVALID SYLLOGISMS

The terms 'Fallacy of Illicit Major', 'Fallacy of Illicit Minor', 'Fallacy of Undistributed Middle' and 'Fallacy of Four Terms' are applied to arguments of traditional syllogistic form which break one or another of a well-known set of rules. We shall describe these in more detail later (in chapter 6).

The Fallacy of Four Terms, however, is worth special mention here because it illustrates a common confusion in modern accounts. An example of a syllogism which is claimed to commit this Fallacy is[1]

> All metals are elements
> Brass is a metal
> Therefore, brass is an element.

[1] Mellone, *Introductory Text Book of Logic*, p. 166.

Here

> the premises have no link of connection, and contain four different terms between them. Such mistakes are possible because of the ambiguity of language. If any term is used ambiguously, it is really *two terms*; hence the syllogism containing it has at least four terms, and is not a true syllogism at all, though at first sight it may appear to be one...
>
> ... using the middle term *metal* in two different senses, in one of which it means the pure simple substances known to chemists as metals, and in the other a mixture of metals commonly called metal in the arts, but known to chemists by the name alloy.

Under this interpretation, the Fallacy of Four Terms is a straightforward case of Equivocation. But this is having things both ways. The middle term cannot be equivocal unless it is *one* term with *two* meanings. If there are really four terms we have a formal fallacy, independently of whether any term is equivocal: if we have an essentially equivocal term there is a fallacy of the Aristotelian variety whatever the formal shape of the argument.

Incidentally, there is nothing to prevent the other terms of a syllogism from being involved in the same kind of trouble. There is nothing to stop us from having a Fallacy of Five Terms, or even a Fallacy of Six Terms.

FALLACIES OF SCIENTIFIC METHOD

Cohen and Nagel have invented special names for a number of fallacies under the heading 'Abuses of Scientific Method'. We read here about the Fallacy of Simplism or Pseudo-Simplicity, and a number of particular fallacies under this heading such as the Fallacy of Exclusive Linearity and the Fallacy of False Opposition. We are told (p. 384) that

> ... hasty monism, the uncritical attempt to bring everything under one principle or category, is one of the most frequent perversions of scientific method.

We also hear of the Genetic Fallacy, which is the fallacy of confusing temporal or historical origin with logical nature. There is no attempt at system or at completeness.

The notion that invalid induction is a species of fallacy is first explicit in the *Port Royal Logic* (p. 264):

> Finally, we reason sophistically when we draw a general conclusion from an incomplete induction. When from the examination of many particular instances we conclude to a general statement, we have made an induction. After the waters of many seas have been found salty and the waters of many rivers found fresh, we can conclude that sea water is salty but river water is fresh. . . . It is enough to say here that imperfect inductions – that is, inductions based on examination of fewer than all instances – often lead us to error.

The authors go on to give the example of the generalization 'Suction pumps can raise water to any height', and the (at that time recent) discovery that there is in fact a height limit of about thirty-two feet. We shall look later at the historical ancestry of this concept of fallacy. It has had few descendants but there has been a thread of interest in it, and one of its main exponents is J. S. Mill (*System of Logic*, Book V). Of the modern books we have reviewed, only one, Salmon, has an explicit treatment of inductive shortcomings. Here we read (p. 56)

> The *fallacy of insufficient statistics* is the fallacy of making an inductive generalization before enough data have been accumulated to warrant the generalization. It might well be called 'the fallacy of jumping to a conclusion.'

We have met Hasty Generalization before under the heading of the Fallacy of *Secundum Quid:* though not in Salmon. Again (p. 57)

> The *fallacy of biased statistics* consists of basing an inductive generalization upon a sample which is known to be unrepresentative or one which there is good reason to believe may be unrepresentative.

A traditional example of this is from Francis Bacon, who gives a psychological explanation:[1]

> . . . that instance which is the root of all superstition, namely, *That to the nature of the mind of all men it is consonant for the affirmative or*

---

[1] *Advancement of Learning* (1605); see *Works*, vol. 3, p. 395. The example is originally from Cicero, *Nature of the Gods*, III, § 89.

active to affect more than the negative or privative: so that a few times hitting or presence, countervails oft-times failing or absence; as was well answered by Diagoras to him that shewed him in Neptune's temple the great number of pictures of such as had scaped shipwrack and had paid their vows to Neptune, saying, *Advise now, you that think it folly to invocate Neptune in tempest: Yea but* (saith Diagoras) *where are they painted that are drowned?*

Joseph (p. 595) gives this as an example of False Cause. It will be seen that it is not very difficult to find a place in the traditional scheme for such examples, and this apparently tends to discourage go-it-alone originality of Salmon's variety.

The difficulty that surrounds the definition of 'inductive fallacies' in their own right is that of distinguishing at all precisely between good inductions and bad. Every Philosophy student knows what Hume made of this difficulty. What is most clearly wrong with the *Port Royal* account and its successors is that nothing definite is done to provide criteria. If an induction is based on 'fewer than all instances' it may lead us into error, but it also may not. This leaves it open to everyone to adopt any inductive argument that happens to please him and to censure as fallacious any he happens to dislike. To add to the public confusion, logicians are in the habit of presenting induction as an argument from particular to general in such a way as to guarantee that it commits the Fallacy of the Consequent. In one chapter of a textbook we are shown schemata such as

Crow No. 1 is black
Crow No. 2 is black
. . .
Crow No. n is black
Therefore, all crows are black.

as examples of conditionally valid inductions and in another we are given comparable arguments as unconditionally fallacious. We have already noticed the tendency to give as examples of the Fallacy of the Consequent instances of what might be construed as valid, or at least incipient, inductions.

Once again, what is needed is some logical clarification. Until it is clear whether induction is an argument-form in any way comparable with deduction there is nothing to be gained by treating inductive shortcomings as varieties of fallacy.

MISCELLANEOUS

There are several varieties of fallacy or particular Fallacies which have received special names, but which are not really logical fallacies at all but merely false beliefs. This is the sense in which the word 'fallacy' is used in the title of Martin Gardner's book *Fads and Fallacies in the Name of Science*, subtitled *A Study in Human Gullibility*, which examines for us every kind of scientific crankhood from water divining to scientology. In more philosophical contexts names for particular erroneous or allegedly erroneous doctrines have often been invented and have sometimes received currency. The *pathetic fallacy*, for example, is the mistake of supposing that nature and inanimate objects have feelings like humans; and infects, at least as a literary device, certain kinds of tragic writing. Most famous in modern times, perhaps, is G. E. Moore's *naturalistic fallacy*, which he describes as follows (*Principia Ethica*, p. 10):

> It may be true that all things which are good are *also* something else, just as it is true that all things which are yellow produce a certain kind of vibration in the light.... But far too many philosophers have thought that when they named those other properties they were actually defining good; that these properties, in fact, were simply not 'other', but absolutely and entirely the same with goodness. This view I propose to call the 'naturalistic fallacy'...

He is concerned with the identification of moral values with pleasure, usefulness, majority approval, 'greatest happiness of the greatest number', or any other quality which might conceivably be described independently of morality. In these cases, Moore thinks, we can still ask 'It is pleasant (useful, etc.) but is it good?', illustrating that there is not an identity between the terms.

Different classificatory systems for the logical fallacies lead to the invention of different classificatory terms. J. S. Mill, for example, though his account (*System of Logic*, Book V) has little in it that is not covered in essence in what has already been said, has an original scheme of classification involving five broad categories: (1) Fallacies *a priori*, which are false beliefs, prejudices or superstitions with which people approach a subject-matter;

(2) Fallacies of Observation, where the subject-matter itself is falsified; (3) Fallacies of Generalization, including his treatment of faulty induction and false analogies; (4) Fallacies of Ratiocination, which are formal; and (5) Fallacies of Confusion. Mill took this last title from Bentham, and it represents his ragbag category, including Question-Begging, Ambiguity, and *Ignoratio Elenchi*.

We shall have occasion later to look at some other attempts at reclassification. Most modern writers have their minor preferences of arrangement, but it is almost always the same material that is being chopped about and served up reheated. One has the impression that respect for the material or the tradition has long since disappeared; and the great argument for conformity is that it saves effort. For the last word on this subject I can do no better than quote from the influential seventeenth-century *Compendium* of Aldrich. The book is in Latin; and the section on fallacies, which contains no novelty of treatment whatever, ends as follows (Appendix §4):

> These, then, are the thirteen kinds of fallacy familiar to the ancients and normally presented to Logic students as examples. The number could be cut down; for some seem to coincide and, moreover, three of them – Non-Cause, Begging the Question and Many Questions – are not fallacies properly so-called, that is, ill-formed syllogisms, but rather faults of the opponent. The number could also be increased; but since it satisfied Aristotle it has satisfied all later logicians.

CHAPTER 2

Aristotle's List

The tradition described in the previous chapter is so incoherent that we have every reason to look for some enlightenment at its historical source. Since the division of fallacies found in the modern books is, in the main, a development of that of Aristotle in his *Sophistical Refutations*, it is to Aristotle that we must turn.

We need do no more than open a copy of the *Sophistical Refutations* to find features inconsistent with modern conceptions. To start with, even the title presents us with some questions: why 'refutations'? A *refutation* (Greek ἔλεγχος, *elenchus*) is defined near the start of the treatise as 'reasoning involving the contradictory of the given conclusion'; but though this might reassure us as to the correctness of the translation it adds to the mystery since it is not clear why we must assume that we are presented with a 'given conclusion' or why it is only or mainly reasoning *against* such a conclusion that is going to be investigated. Again, the word 'sophistical' seems to have a shade of meaning that is not quite appropriate. Greek has no precise synonym for 'fallacy', and the word so translated is generally σόφισμα, *sophisma*; but to discover this is no comfort.

That Aristotle is writing about *deliberate* sophistry, and not about mere errors or mistakes, is made clear quite early (165a 19):

> Now for some people it is better worth while to seem to be wise, than to be wise without seeming to be (for the art of the sophist is the semblance of wisdom without the reality, and the sophist is one who makes money from an apparent but unreal wisdom); ...
> Those, then, who would be sophists are bound to study the class of arguments aforesaid: ...

Among the book's aims, then, is even the training of would-be 'sophists' in the use of fallacious arguments, so that they may deceive others and make money in so doing! To be sure, there is also the subsidiary aim of rendering the reader proof against deception; but this is also only a subsidiary aim in the modern books.

The next thing that might worry us is the unexplained occurrence of references to 'questioning' and 'the questioner', 'answering' and 'the answerer'. Are we being treated to training in cross-examination in courts of law? Actually what is referred to is a particular form of academic or public debate, and we shall look at details in a moment. The scope of the book's investigation is laid out as follows (165b 12):

> First we must grasp the number of aims entertained by those who *argue as competitors and rivals to the death*. These are five in number, refutation, fallacy, paradox, solecism, and fifthly to reduce the opponent in the discussion to babbling – i.e. to constrain him to repeat himself a number of times: *or it is to produce the appearance of each of these things without the reality.*

(My italics.) Some of these unscholarly aims, such as 'solecism' ('to make the answerer, in consequence of the argument, to use an ungrammatical expression') and the 'reduction of the opponent to babbling' turn out, when fully described, to be less vexatious than they sound; but this is, nevertheless, the general tenor of the treatise. Opponents in debate should, if possible, be plainly refuted; but if this is not possible they should either be convicted of *some* falsehood, relevant or irrelevant; or led into paradox; or reduced to solecism or babbling; or, if none of these aims can be achieved by 'fair' means, they should be achieved by foul, by producing the appearance without the reality.

Later on Aristotle devotes a whole chapter to *stratagems* of argument. Thus (174a 16)

> With a view then to refutation, one resource is length – for it is difficult to keep several things in view at once; and to secure length the elementary rules that have been stated before should be employed. One resource, on the other hand, is speed; for when people are left behind they look ahead less. Moreover, there is anger and contentiousness, for when agitated everybody is less

able to take care of himself. Elementary rules for producing anger are to make a show of the wish to play foul, and to be altogether shameless.

Or again (174b 7, 38)

> A strong appearance of having been refuted is often produced by the most highly sophistical of all the unfair tricks of questioners, when without proving anything, instead of putting their final proposition as a question, they state it as a conclusion, as though they had proved that 'Therefore so-and-so is not true'...
>
> ... One must not ask one's conclusion in the form of a premiss, while some conclusions should not even be put as questions at all; one should take and use it as granted.

There is nothing quite like this anywhere in modern literature.[1]

These quotations are enough to raise serious questions about the traditional interpretation of Aristotle's list of fallacies. Nevertheless there is something to be said on the other side; and, to do so, we shall have to put the *Sophistical Refutations* in its historical setting and see it against a background of Greek academic institutions and the development of Greek thought.

The key to the understanding of the work is that it was composed while Aristotle was still a young man, a student-teacher in Plato's Academy; before his epoch-making work on Formal Logic and, therefore, before Formal Logic had been invented. It has too often been regarded as appendical to a manual of formal reasoning. For historical perspective we must see it as itself a Logic-in-the-making that is quite different in character.

Aristotle wrote three accounts of fallacies, spread through his career. The *Sophistical Refutations* is the earliest and shows the strong influence of some of Plato's dialogues. The other accounts are in the *Prior Analytics* and the *Rhetoric*. Aristotle's active life – though there is hardly a fact about it or about the dating of his works that is not contested – is generally accepted as falling into

---

[1] I hope not to be accused of frivolity in suggesting that the closest parallel is in Stephen Potter's *Gamesmanship, Lifemanship* and other writings in this vein. Interesting parallel passages can be found. It is no disparagement of these works to suggest that their success as satire would not have been possible if the serious study of disputation had not been neglected in modern times.

three main periods.[1] He first went to the Academy about 367 B.C. at the age of seventeen and stayed as student and teacher for twenty years. He might reasonably have expected to succeed to leadership of the Academy; but when, on Plato's death in 347, the post went elsewhere he left, married soon after, and during the next twelve years lived at several different places in the Greek islands and Asia Minor and was latterly tutor to the young Alexander the Great at the court at Pella in Macedon. When Alexander succeeded to the throne in 335 he returned to Athens and for a further period of about twelve years conducted his own school known as the Lyceum.

Nobody is quite sure when the *Prior Analytics* was written but a reasonable guess might put it towards the end of his stay in the Academy or, perhaps, during a relatively quieter regime of research and domestic bliss shortly after he left. This is the work which presents Aristotle's theory of deduction and the syllogism and which can be regarded as originating Formal Logic. The notes on fallacies are brief, but interesting in their selectivity and arrangement, and indicate a development of his thought since the *Sophistical Refutations*.

The *Rhetoric* (1390b 10) contains the interesting statement that the mind is in its prime about the age of forty-nine, and this is almost enough on its own to permit us to date it in 335, the year of his return to Athens: other internal evidence is consistent with this, at least as regards the first two of its three books. However, he had been thinking about and teaching Rhetoric for many years before. Both the theory of inference and its complementary discussion of fallacies are adapted to the special needs of the subject.

There is one complication to this picture of composition-dates. It is clear from the patchwork nature of some of Aristotle's works that they have been extensively edited and rearranged since they were first written. In particular, cross-references have been inserted, often appearing to indicate composition in an order corresponding with the modern order of arrangement and which appeared to the editor as 'logical'. It is not clear whether this was done by Aristotle himself during his second period in Athens, or by Theophrastus or some other successor at the

[1] See Jaeger, *Aristotle*; Ross, *Aristotle*; Grene, *A Portrait of Aristotle*.

Lyceum, or even much later in the first century B.C. when Andronicus of Rhodes is known to have prepared the works for republication. The most reasonable speculation seems to be that Aristotle revised his own material for use in lectures at the Lyceum.

When Aristotle first came to Athens Plato was sixty-one, an influential philosopher who was just entering his period of most mature work as reflected in the dialogues *Theaetetus*, *Sophist*, *Statesman*, *Parmenides*, *Philebus*, and *Timaeus*. The material of these works was Aristotle's philosophical training, and at first he even emulated Plato by writing dialogues of his own. Philosophy, in Plato's Academy, must have been studied very much in dialogue form. Not only were there formal debates and contests, but even in the classroom a good deal of the instruction must have been, as often in modern tutorials, in the form of a discussion between the teacher and one or another selected pupil. Plato had a theory that all knowledge worth the name is already, strictly speaking, in the possession of the learner and only needs to be elicited by means of suitable questions. The questioner's role, says Socrates in the *Meno*, is that of a midwife. Sometimes in the dialogues a speaker (usually Socrates), having a didactic point to make, deliberately selects a member of the audience and plies him with questions.

Writing, in his *Topics* and *Sophistical Refutations*, on arguments in dialogue form, Aristotle especially distinguishes a class of arguments that he calls 'didactic'; they are (165b 1)

> ... those that reason from the principles appropriate to each subject and not from the opinions held by the answerer (for the learner should take things on trust):...

For Plato and Aristotle a dialogue, didactic or not, is always between two people identified as 'the questioner' and 'the answerer'. Plato might not have agreed that the learner should take things on trust, since he thinks of the educative process as one of a clash of opinions rather than as a serving-up of pre-systematized knowledge. However, in his more didactic works such as the *Republic*, much of the dialogue is of a rather one-sided character, where the answerer's principal contributions are indications of assent such as 'Assuredly' or 'That is certainly so,

Socrates'. Elsewhere the answerer humbly raises difficulties that appear to him to require further comment. Later Aristotle apparently dropped this kind of tuition and replaced it with formal lectures; but there is no firm evidence that Plato or anyone else gave formal lectures in the Academy.

Didactic arguments, however, are not the only kind of argument in dialogue form. Aristotle distinguishes at least two other kinds, *dialectical* or *examination* arguments, and *contentious* arguments. The last named are (165b 12) 'the arguments used in competitions and contests', and the *Sophistical Refutations* is intended, according to its opening paragraphs, to be relevant particularly to these. There is some evidence that public debates, as much for entertainment as for instruction, may have been a feature of life in Athens. We meet professional debates called 'sophists' who make a practice of taking part in contests and train students to do the same. The better-known sophists such as Gorgias and Protagoras are worthy philosophical foils for Socrates, and Plato's dialogues named after them relate encounters that are more than mere debating competitions. Both Plato and Aristotle, however, have a low opinion of sophists and have bequeathed us the word 'sophistical' in a meaning that reflects their assessment. Aristotle says (*Topics* 100b 23)

> reasoning is 'contentious' if it starts from opinions that seem to be generally accepted, but are not really such, or again if it merely seems to reason from opinions that are or seem to be generally accepted.

A contentious inference is called a *sophism* (162a 17), so that it is written into the definition that the stock-in-trade of sophists is fallacy.

We have no detailed account of an actual sophistical contest. The best example, in Plato's dialogues, of the kind of argumentation he and Aristotle have in mind is the *Euthydemus*. This dialogue seems to have impressed Aristotle, since he not only takes a number of examples from it in his *Sophistical Refutations* but also reproduces the mood of Socrates's reproof of the sophists and praise of Philosophy in his *Protrepticus*. It is, however, one of Plato's earlier dialogues. Two sophists, Euthydemus and Dionysodorus, in turn practice their craft on the young boy

Cleinias, tying him in logical knots. Socrates is relating the encounter (*Euthydemus* 275):

> Now Euthydemus, if I remember rightly, began nearly as follows: O Cleinias, are those who learn the wise or the ignorant?
>
> The youth, overpowered by the question, blushed, and in his perplexity looked at me for help; . . . Whichever he answers, said Dionysodorus, leaning forward so as to catch my ear, his face beaming with laughter, I prophesy that he will be refuted, Socrates.
>
> While he was speaking to me, Cleinias gave his answer: . . . that those who learned were the wise.
>
> Euthydemus proceeded: There are some whom you would call teachers, are there not?
>
> The boy assented.
>
> And they are the teachers of those who learn — the grammar-master and the lyre-master used to teach you and other boys; and you were the learners?
>
> Yes.
>
> And when you were learners you did not as yet know the things which you were learning?
>
> No, he said.
>
> And were you wise then?
>
> No, indeed, he said.
>
> But if you were not wise you were unlearned?
>
> Certainly.
>
> You then, learning what you did not know, were unlearned when you were learning?
>
> The youth nodded assent.
>
> Then the unlearned learn, and not the wise, Cleinias, as you imagine.
>
> At these words the followers of Euthydemus, of whom I spoke, like a chorus at the bidding of their director, laughed and cheered. Then, before the youth had time to recover his breath, Dionysodorus cleverly took him in hand and said: Yes, Cleinias; and when the grammar-master dictated anything to you, were they the wise boys or the unlearned who learned the dictation?
>
> The wise, replied Cleinias.

> Then after all the wise are the learners and not the unlearned; and your last answer to Euthydemus was wrong.
>
> Then once more the admirers of the two heroes, in an ecstasy at their wisdom, gave vent to another peal of laughter, ...

Socrates intervenes on behalf of Cleinias and the sophists turn their attention to him (283):

> Dionysodorus said: Reflect, Socrates; you may have to deny your words.
>
> I have reflected, I said, and I shall never deny my words.
>
> Well, said he, and so you say that you wish Cleinias to become wise?
>
> Undoubtedly.
>
> And he is not wise as yet?
>
> At least his modesty will not allow him to say that he is.
>
> You wish him, he said, to become wise and not to be ignorant?
>
> That we do.
>
> You wish him to be what he is not, and no longer to be what he is?
>
> I was thrown into consternation at this.
>
> Taking advantage of my consternation he added: You wish him no longer to be what he is, which can only mean that you wish him to perish. Pretty lovers and friends they must be who want their favourite not to be, or to perish!

In the course of the dialogue the sophists prove, among other things, that it is impossible to tell a lie, since a lie is 'that which is not', and can have no existence; that good men speak evil, since they would not be good if they did not speak evil of evil things; that no one ever contradicts anyone; that Socrates both knows (some things) and does not know (other things) at the same time; and that everything visible 'has the quality of vision' and hence can see. Towards the end of the dialogue the quibbles run riot, with Socrates and another onlooker Ctesippus joining in. Here we have (298):

> If you will answer my questions, said Dionysodorus, I will soon extract the same admissions from you, Ctesippus. You say that you have a dog.
>
> Yes, a villain of a one, said Ctesippus.

And he has puppies?

Yes, and they are very like himself.

And the dog is the father of them?

Yes, he said, I certainly saw him and the mother of the puppies come together.

And is he not yours?

To be sure he is.

Then he is a father, and he is yours; ergo, he is your father, and the puppies are your brothers.

Let me ask you one little question more, said Dionysodorus, quickly interposing, in order that Ctesippus might not get in his word: You beat this dog?

Ctessipus said, laughing, Indeed I do; and I only wish that I could beat you instead of him.

Then you beat your father, he said.

We have already met this example.

Plato's attitude to these arguments is mixed. He puts heavy irony into Socrates's praise of them (303-4):

> There is much, indeed, to admire in your words, Euthydemus and Dionysodorus, but there is nothing that I admire more than your magnanimous disregard of any opinion – whether of the many, or of the grave and reverend seigniors – you regard only those who are like yourselves ...
>
> ... but at the same time I would advise you not to have any more public entertainments; there is a danger that men may undervalue an art which they have so easy an opportunity of acquiring; the exhibition would be best of all, if the discussion were confined to your two selves; but if there must be an audience, let him only be present who is willing to pay a handsome fee: ...

Crito, to whom Socrates is relating the encounter, agrees that he 'would rather be refuted by such arguments than use them in the refutation of others'. Nevertheless Plato approves of the question-and-answer procedure and is genuinely fascinated by the paradoxes. Among these there is a least one – the contention that if a beautiful thing is not identical with absolute beauty it cannot really be beautiful – that Plato would want to espouse as his own. Even the sophists' disregard of the opinions of others is something that Plato, in other moods, would be prepared to approve without irony. With hindsight we can see the *Euthydemus* as an exercise in Logic, and the absurdities of the sophists as a set of puzzles for the would-be theoretical logician. Aristotle's *Sophistical Refutations* can then be regarded as a first step in constructing the relevant logical theory.

How, then, should arguments be conducted? Let us turn to *dialectical* or *examination* arguments. Aristotle is not quite sure whether these are two kinds or one, and in different places makes each a subclass of the other (*Soph. Ref.* 169b 25, 171b 4, 172a 23). The word 'dialectical', which seems to have been invented by Plato, is partly honorific in force and refers to all that is best in Plato's way of doing Philosophy. Aristotle says of his theory of argumentation in the *Topics* (101a 34–b 4)

> ... the ability to raise searching difficulties on both sides of a subject will make us detect more easily the truth and error about the several points that arise. ... Dialectic is a process of criticism wherein lies the path to the principles of all inquiries.

The contrast between the three kinds of argument – didactic, dialectical or examination, and contentious – is clearly drawn in the long middle clause of the following passage (*Topics* 159a 25):

> Inasmuch as no rules are laid down for those who argue for the sake of training and of examination: – and the aim of those engaged in teaching or learning is quite different from that of those engaged in a competition; as is the latter from that of those who discuss things together in a spirit of inquiry: for a learner should always state what he thinks: for no one is even trying to teach him what is false; whereas in a competition the business of a questioner is to appear by all means to produce an effect upon the other, while that of the answerer is to appear unaffected by him; on the other hand, in an assembly of disputants discussing in the spirit not of a competition but of an examination and inquiry, there are as yet no articulate rules about what the answerer should aim at, and what kind of things he should and should not grant for the correct or incorrect defence of his position: – inasmuch, then, as we have no tradition bequeathed to us by others, let us try to say something upon the matter for ourselves.

Dialectical or examination arguments are those conducted 'in a spirit of inquiry' and Aristotle *Topics* – or at least this part of it, Book VIII – is to be a manual for their conduct. At the same time, much of the rest of the *Topics* also refers in spirit to these arguments; and the same is true of the *Sophistical Refutations*, which follows on from it and has sometimes been regarded as its Book IX.

A dialectician, says Aristotle (*Soph. Ref.* 172a 35), is a man who examines with the help of a theory of reasoning. The skill of the speakers in Plato's most mature dialogues, which are his model, is the result of a long training in move and countermove. Good discussions must have experts on both sides. On the other hand, there is no easy way of characterizing expertness and Plato sometimes seems to say that it is impossible to do so:[1] there is no set of rules but only inspiration, a divine gift. Aristotle is not satisfied to leave it at this but his actual definition of dialectical arguments is less than satisfactory (*Topics* 100a 30, b 21):

> ... reasoning, on the other hand, is 'dialectical', if it reasons from opinions that are generally accepted ...

> ... those opinions are 'generally accepted' which are accepted by everyone or by the majority or by the philosophers – i.e. by all, or by the majority, or by the most notable and illustrious of them.

This marks them off from didactic arguments and, as defined above, contentious arguments but does not give us any clue to their supposed exceptional merit.

In fact, Aristotle is in transition from a pure Platonic view to a more measured one which treats Dialectic as a mere technique, unessential to the pursuit of truth. At times – see, for example *Soph. Ref.* 169a 37 – he even thinks of it as a hindrance: he is in the process of discovering Logic which, he thinks, enables a man to achieve as much by solitary thought as in social intercourse. But Aristotle is the great systematizer, and the first system he must build is one which will display and label the various parts and species of the Method (the word is also Plato's: see, for example, *Republic* 533) which has taught him so much. Already, in his old age, Plato had been growing away from Dialectic. Aristotle completes the process, by the paradoxical procedure of

---

[1] For example, at *Republic* 533. For a discussion of Plato's concept see Robinson, *Plato's Earlier Dialectic*, pp. 69 ff.

consummating a theoretical work on the dialectical method. This is a contradiction in terms, and the dialectical method is dead. Aristotle is half aware of this himself (see *Rhetoric* 1359b 12); and he is conscious that he is breaking new ground. He contrasts the subject with that of Rhetoric, and says (*Soph. Ref.* 183b 34)

> Of this inquiry [Dialectic], on the other hand, it was not the case that part of the work had been thoroughly done before, while part had not. Nothing existed at all. For the training given by the paid professors of contentious arguments was like the treatment of the matter by Gorgias. For they used to hand out speeches to be learned by heart, some rhetorical, others in the form of question and answer, each side supposing that their arguments on either side generally fall among them.

From various works we can dig out some notion of the rules to which a debate was supposed to conform.[1] Aristotle generally assumes that the same rules apply to debates of all kinds. The aim of a debate is generally understood to be the discovery of the nature or essence of some abstract concept, e.g. virtue, or number, or the State; but there were various standard or recurring subjects, of which we possess lists: many of these were current in Greek philosophy long before Plato. There is one questioner and one answerer at a time. The answerer commits himself to defending a general thesis on some subject, and submits to questioning. The questions must be clear and direct – in fact, of the kind known as 'leading questions' – and, if they are not, the answerer may object to them and insist that they be clarified or reformulated. The answerer is expected to give his own opinion, independently of the opinions of the questioner or of others; except of course that, if he is found out in a contradiction, he must change one or other of the contradictory opinions in order to escape it. He is expected to have an opinion in answer to each properly formulated question; but, if he must profess simple ignorance or lack of opinion on any point, it becomes the business of the questioner to formulate further questions which will lead him to a resolution of it. Every issue must be followed through: the questioner and answerer may not 'agree to differ'. The dialogue or phase of dialogue ends when the questioner has

---

[1] See particularly Robinson, *Plato's Earlier Dialectic*, pp. 69 ff.

refuted the answerer or when it becomes clear that he cannot do so. Bystanders or audience are understood to know these rules and to be capable of intervening to enforce them if necessary, and to be the ultimate adjudicators. Ryle thinks (*Plato's Progress*, ch. 4) that both impromptu and staged dialogues were common from before the time of Socrates, that is, from the fifth century; that some of Plato's dialogues were written for public performance at the occasion of Olympic games; and that Plato and Aristotle themselves took part in these performances.

About three-quarters of Aristotle's *Topics* consists of rules of argumentation which can be construed as a groping attempt at a formal theory of inference. A *topic* is a form of argument: in Latin this became *locus* or *locus communis* – Aristotle has both 'special' and 'common' topics – whence the English word *commonplace* which, in logical as distinct from rhetorical tradition, used to refer to such a rule or argument-form. Book VIII is explicitly concerned with the practice, as against the theory, of Dialectic and includes, in particular, rules for putting questions and for answering them.

The *Sophistical Refutations* explicitly takes up the question of 'contentious' reasoning. It falls rather clearly into two parts which were later, in the medieval vulgate translation, called Book I and Book II[1] under the subtitles *Invention* and *Judgment*. The first part describes how various 'sophistical refutations' arise, and the second discusses their solution and avoidance. In each part the same thirteen sophistical refutations are enumerated, divided into their two groups as follows:

[a] Sophistical refutations dependent on language.

[1] Ambiguity.
[2] Amphiboly.
[3] Combination of words.
[4] Division of words.
[5] Wrong accent.
[6] The form of expression used.

[b] Sophistical refutations not dependent on language.

[1] Accident.
[2] The use of words absolutely or in a certain respect.

[1] As in Migne, vol. 64, cols. 1007–40.

[3] Misconception of refutation.
[4] Assumption of the original point.
[5] Consequent.
[6] Non-cause as cause.
[7] Making of two questions into one.

That we are not concerned with fallacies of reasoning in quite the modern sense is made clear from the structure or architectonic within which the discussion is embedded. To start with, there is a third category:

[c] Sophistical refutation by valid arguments inappropriate to the subject-matter.

Under this heading Aristotle describes what is usually presented under [b] 3 as *Ignoratio Elenchi*: we shall see later how he really conceives [b] 3.

Secondly, however, the three categories of sophistical refutations together only make up the first of the five kinds of contentious reasoning mentioned earlier. These are distinguished as five possible aims of the questioner, in detail as follows:

A. To refute the opponent; that is, to prove the contradictory of his thesis.
B. To show that the opponent has committed a fallacy.
C. To lead the opponent into paradox.
D. To make the opponent use an ungrammatical expression.
E. To reduce the opponent to babbling; that is, make him repeat himself.

The work opens with a definition of 'refutation', some remarks about sophists and their practices, and a classification of arguments in dialogue form – in this place, into four categories, dialectical and examination arguments being regarded as distinct. The five possible aims of the questioner in a contentious argument are listed. Taking the first of these, refutation, the thirteen traditional kinds in their two groups are introduced and exemplified in order. Having worked through his list Aristotle next suggests that all 'apparent proofs and refutations' may be referred to 'ignorance of what refutation is', that is, to category [b] 3, and goes through them again briefly to prove his point.

There follows a puzzling short chapter whose main point seems to be the psychological one that it is very easy to be misled by these sophisms if we do not pay due attention to fine logical distinctions.

Next the third kind of refutation – by arguments which, though valid, are inappropriate to the question-at-issue – is introduced rather as an afterthought. There follows a collection of short discussions on the nature of sophistical refutation in general: it is never absolute but only relative to a particular answerer: there is really an infinity of kinds of refutation and we cannot study them all: our concern is with those common to all the sciences and not with those peculiar to a particular science, such as Geometry. There is an interesting discussion of the distinction which 'some people draw' – one sees Aristotle arguing against a possible interpretation of his own twofold classification – between 'arguments that are directed against the expression' and 'arguments that are directed against the thought expressed'; but this is ultimately only a distinction between cases in which the expression is, and is not, used in the same sense by questioner and answerer. Some attempt is made to elucidate the finer shades of meaning of 'contentious' and 'sophistical' and to contrast professional sophists with 'quarrelsome persons' who argue contentiously 'in order to win the mere victory'. One of the things that distinguishes contentious reasoning as Aristotle studies it, moreover, is that it is common to all subjects of inquiry, and this especially unfits it for use in the special sciences since (172a 8):

> ... an argument which denied that it was better to take a walk after dinner, because of Zeno's argument, would not be a proper argument for a doctor, because Zeno's argument is of general application.

(The relevant argument of Zeno is that no one can make a journey since he must first make half of it, and before that, half, and so on: see *Physics* 263a 4) Everybody, even including amateurs, makes use in a way of Dialectic.

Aims B, C, D, and E of the contentious reasoner are now considered: we shall return to some of these later. There follows the chapter on stratagems of a general nature: length, speed and so on.

Turning, in what the Middle Ages reckoned as Book II of the treatise (175a 1) to 'answering, and how solutions should be made', we are given a paragraph on the philosophical use of the study of sophisms: if we can learn to detect them we shall be helped towards avoiding them in our own thought. Here practice and training in spotting is as important as theoretical labelling. However, Aristotle brings us quickly back to the exigencies of disputation: it is as important for the answerer to use sophistical tricks as it is for the questioner since, quite independently of the merits of the argument, it is important for him to avoid the *appearance* of being refuted. So that the correct technique of answering may be demonstrated, he runs through the entire list of sophistical refutations in detail again, in a treatment which often adds to our understanding of what may have been too briefly said earlier. At the end of the discussion of refutations dependent on language – category [*a*] – a section is inserted giving a number of additional examples and containing an analysis of Plato's 'third man' argument (compare Plato, *Parmenides*, pp. 130–2) which is to the effect that besides individual men and the general concept 'man' there is a third concept that can be predicated of both the first two: Aristotle's well-known solution is that the general concept 'man' denotes 'not an individual substance, but a particular quality, or the being related to something in a particular manner, or something of that sort', whence the fault lies in a sophism dependent on the form of the expression, category [*a*] 6. After all the thirteen sophisms of categories [*a*] and [*b*] have been dealt with from this point of view there is a similar discussion of arguments aimed at 'babbling' and solecism and another chapter discussing the relative psychological force of different argument-forms. The treatise ends with an epilogue which sums up the programme of the *Topics* as well as of the *Sophistical Refutations* and compares the state of the theory of Dialectic with that of other subjects, particularly Rhetoric.

Aristotle, like Plato, is at once contemptuous of sophistical reasoning and fascinated by it. He is not always sure how to distinguish good reasoning from bad and is prepared, now and then, to appeal to irrelevancies such as the fact that sophists practise for money. Dialectic is standing in the way of his understanding of Logic. For ourselves, a move in the other direction

E

has been indicated: in our attempts to understand Aristotle's account of fallacies we need to give up our tendency to see them as purely logical and see them instead as moves in the presentation of a contentious argument by one person to another. Another moral we might draw is that the explicit study of Dialectic might have something to offer the modern student.

THE ACCOUNT IN THE 'PRIOR ANALYTICS'

Bochenski (*History of Formal Logic*, p. 55) notes:

> There is a second doctrine of fallacious inference, in the *Prior Analytics*, much briefer than the first but incomparably more formal; all fallacious inferences are there reduced to breaches of syllogistic laws.

I do not think this is quite apt. In the first place, there is little in the *Prior Analytics* that is really new and, secondly, what there is contains no definition or clear doctrine: at most, there is a change of direction. By studying this account, however, we might hope to answer certain questions. Did Aristotle, having invented Formal Logic, change his conception and classification of fallacies? Have the writers who have restewed the *Sophistical Refutations* all these centuries missed a change in the direction of his thought? If Aristotle had worked out his new approach more completely, might we have been saved two millennia of confusion?

Perhaps even the fact that we do not have a long or fully-fledged account is significant. Sophistical refutations, as they had been conceived, should have been left behind altogether with Dialectic. But Aristotle never quite succeeded in leaving them behind for, when he turns to the discussion of fallacy, he tends also to turn again to the irrelevant concepts of questioning and answering and to assume again that errors of reasoning are the result of deliberate deceit. Even later, in the *Rhetoric*, there are still undertones of dialogue, for the rhetorical practitioner is thought of as drawing out an argument in his audience as a questioner might.

A brief and obviously late-interpolated passage at the beginning of the *Sophistical Refutations* refers to the *Prior Analytics* as providing a theoretical treatment of didactic argument (165b 8).

This identification is intended to make us see the whole of the *Organon* – *Categories, Interpretation, Prior Analytics, Posterior Analytics, Topics* and *Sophistical Refutations*, in that order – as a unity, in which arguments of all three (or four) kinds are discussed. This is also the rationale of this traditional arrangement of the works: the *Prior Analytics* is supposed to deal with didactic reasoning, the *Topics* with dialectical and examination arguments, and the *Sophistical Refutations* with contentious. However, the *Prior Analytics* defies this classification and is nearly all devoted to a formal and context-free appraisal of argument-forms. The word 'syllogism', which appears in the *Topics* but is used there in reference to any kind of verbal proof, in the *Prior Analytics* acquires by association the specialized meaning it still has, in reference to a three-term three-proposition argument based on the four standard forms of proposition and their modal counterparts in the shape 'If all men are mortal and all Greeks are men, all Greeks are mortal', or 'If bus services are necessarily inefficient and some city transport facilities are bus services, it is possible that not all city transport facilities are efficient'. Aristotle gives a complete theory of deduction for these propositions, surprisingly modern in its thoroughness. The reduction of all categorical syllogisms to the four 'perfect' forms of the first figure has the nature of an axiomatization.[1] Nowhere, however – even in the supplementary discussions on such subjects as how to find arguments in support of a given conclusion – is there any necessity for specification of the context of use of the rules that are given. This completely marks the *Prior Analytics* off from the other works.

The part of the *Prior Analytics* that particularly concerns us is Book II, chapters 16–21 where, having disposed of the rules for validity of syllogisms, Aristotle turns to defective proofs (64b 28):

> To beg and assume the original question is a species of *failure to demonstrate* the problem proposed; . . .

(My italics.) We are immediately treated to examples of other ways in which *failure in demonstration* may occur:

---

[1] See Łukasiewicz, *Aristotle's Syllogistic*, chapters 3 and 4.

A man may not reason syllogistically at all, or he may argue from premisses which are less known or equally unknown, or he may establish the antecedent by means of its consequents; for demonstration proceeds from what is more certain and is prior.

There is an echo here of some of the sophistical refutations independent of language, of the earlier work. Besides Begging the Question we detect Misconception of Refutation and Consequent, and later Non-Cause is discussed. However, there is no mention of questions and answers, refutation, contentious reasoning, or sophists. Begging the Question and Non-Cause are discussed in some detail but interpreted, as we shall see later, entirely formally. An earlier reference to Begging the Question (41b 9) had even been in the context of the pure reasoning involved in the proof of a theorem of Geometry.

There follows a chapter that gives us Aristotle's thoughts about the application of the theory of syllogisms to Dialectic. Here, without warning, we are back in the language of the *Topics*; but the points made are new (66a 25):

> In order to avoid having a syllogism drawn against us, we must take care, whenever an opponent asks us to admit the reason without the conclusions, not to grant him the same term twice over in his premisses, since we know that a syllogism cannot be drawn without a middle term, and that the term which is stated more than once is the middle.

When we are questioning rather than answering, he says, we should conceal the trend of the argument until the last moment by concealing the connection between premisses. It is also a consequence of the theory of syllogisms that an answerer can avoid being refuted if he concedes only negative propositions, or only non-universal ones.

The digression is short-lived and the rest of the *Prior Analytics* is on the other subjects. What are we to say of this as a new treatment of fallacies? In the first place, we should notice that sophisms of the first kind, dependent on language, have been completely dropped, as also have Accident and *Secundum Quid*: all these would be an embarrassment in a formal work where the precise point of formalization is to indicate propositional forms unambiguously. The others of the original thirteen receive at least passing recognition with the exception of Many Questions, which is out of place outside Dialectic. There is evidence from what follows that Aristotle has been reviewing his work on Dialectic to see what changes or new comments he should make. It seems to follow that in reviving the discussion of sophisms he has made a deliberate selection of all the material he considers relevant.

The omissions may strike us as significant; but it does not follow that he has undergone a change of heart. It could be argued, admittedly, that Aristotle was not the man lightly to throw away good material and would have adapted more of it if he could have made it fit. But this assumes that he thought of the *Prior Analytics* as replacing, rather than supplementing, his earlier work and we have no evidence at all for this assumption and a good deal against it. The sureness of his touch in the formal parts of the *Prior Analytics* may even belie his real feelings about this new, untried development; and our modern prejudices need to be carefully discounted as we assess his attitudes and those of his contemporaries. It is too much to expect that Logic, the moment it was invented, should have taken over the field and that Dialectic should immediately have been dropped. There is a case to be argued, even in modern times, on behalf of studies like Dialectic and Rhetoric against a Logic which is pursued in disregard of the context of its use. Logic is an abstraction of features of flesh-and-blood reasoning; and it is entirely natural that a formal theory of fallacies should be seen simply as abstracting features of fallacies drawn in anger. All this suggests that the discussions in the *Prior Analytics* are to be seen not as a new account but as comments supplementary to the old one.

If Aristotle had written a fuller and clearer formal account of fallacies, would it have made any subsequent difference? Historical counterfactuals are notoriously shapeless, but it might just be worthwhile recording the opinion that what the *Prior Analytics* needs at the point at which the discussion occurs is a theory of syllogistic lapses or 'paralogisms' of the kind produced by some later writers. Such a theory, however, leaves the study of sophisms virtually untouched; and does nothing to fill the need this study aims to fill. That Aristotle, at least, did not think that the *Prior Analytics* had superannuated his other work can be demonstrated by turning to consideration of the *Rhetoric*.

THE ACCOUNT IN THE 'RHETORIC'

Aristotle's *Rhetoric* is not usually regarded as a logical work and it tends to be ignored by logicians. Both the classification and the neglect, however, are un-Aristotelian. Some parts of the *Rhetoric*, it is true, are concerned with non-logical questions of style and arrangement of public speeches, and other parts are primarily forensic or political. From the start, however, Aristotle makes it clear that he conceives the subject as a counterpart of Dialectic and as concerned primarily with the presentation of arguments. Thus (1355a 4).

> ... rhetorical study, in its strict sense, is concerned with the modes of persuasion. Persuasion is clearly a sort of demonstration, since we are most fully persuaded when we consider a thing to have been demonstrated.

In the twentieth century we are less sanguine about these matters, but Aristotle is an optimist. Rhetoric is useful (1355a 20)

> ... because things that are true and things that are just have a natural tendency to prevail over their opposites, so that if the decisions of judges are not what they ought to be, the defeat must be due to the speakers themselves, and they must be blamed accordingly.

By 'judges' we must understand not necessarily the judges in a court of law but whoever it is that the arguments are designed to convince; for example, in the case of a political speech, the general public.

Rhetoric has been taught since the dawn of time as training for law or politics. Aristotle still has these applications in mind but defines the study in such a way as to make it relevant to any kind of linguistic presentation, spoken or written. He is reported to have made a collection of earlier training manuals but both he and Plato, whose *Phaedrus* is largely on this subject, are critical of other practitioners on the grounds that they concentrate on style and delivery to the exclusion of content. Aristotle quotes Isocrates, an older contemporary who must have influenced him in his youth, often enough to show that he respects him; but also continually criticizes his 'epideictic' variety of oratory. He virtually ignores the younger, brilliant, Demosthenes.

Rhetoric, in fact, is to be made to model Dialectic quite closely (1356a 35):

> With regard to the persuasion achieved by proof or apparent proof: just as in dialectic there is induction on the one hand and syllogism or apparent syllogism on the other, so it is in rhetoric. The example is an induction, the Enthymeme is a syllogism, and the apparent Enthymeme is an apparent syllogism. I call the Enthymeme a rhetorical syllogism, and the example a rhetorical induction.

The difference between the concepts of Dialectic and the parallel ones of Rhetoric is that the orator has a practical concern to sketch his argument quickly and in a forceful and easily understood form (1357a 16, 1395b 25):

> The Enthymeme must consist of few propositions, fewer often than those which make up the normal syllogism. For if any of these propositions is a familiar fact, there is no need even to mention it; the hearer adds it himself. Thus, to show that Dorieus has been victor in a contest for which the prize is a crown, it is enough to say 'For he has been victor in the Olympic games', without adding 'And in the Olympic games the prize is a crown', a fact which everybody knows.

> ... we must not carry its reasoning too far back, or the length of our argument will cause obscurity: nor must we put in all the steps that lead to our conclusion, or we shall waste words in saying what is manifest.

A further important difference, unremarked but plain to the reader, is that Rhetoric is everywhere less systematic. When lists of 'topics' or lines of argument are given in the *Rhetoric* they are given briefly without detail, unaccompanied by any claim to completeness, and seem to be not so much argument-analyses as mere handy reminders of useful rhetorical ploys. The same applies to the 'topics that form spurious enthymemes' that are the new breed of Fallacies. There are nine of these, namely:

[1] That which arises from the particular words employed:
    (*a*) compact wording giving impression of fresh conclusion;
    (*b*) use of similar words for different things.
[2] To assert of the whole what is true of the parts, or vice versa.

[3] Use of indignant language.
[4] Use of a 'sign', or single instance, as certain evidence.
[5] Representing the accidental as essential.
[6] Argument from consequence.
[7] Representing as causes things which are not causes.
[8] Omitting mention of time and circumstances.
[9] Confusion of absolute with particular.

If we look closely at these we find again a selection from the earlier list of sophisms. In many cases, however, there is hardly more than a family likeness between the old and the new, and it seems that the earlier material has not so much been reproduced as simply used to stimulate new thoughts. Headings [1] and [2] are all that is left of sophistical refutations dependent on language and [2] is a rather strange substitute for Composition and Division: [1] (*a*) was not actually in the original list at all, but mentioned separately (see 174b 7). [3] and [4] are new and are specially relevant to Rhetoric, the latter an important kind of inadequate enthymeme. [5], [6] and [7] correspond closely with three of the sophistical refutations independent of language, and [8] and [9] seem both to be cases of *Secundum Quid*, the old [*b*] 2. There is no mention of Begging the Question, or Many Questions, though there is no reason to regard these as irrelevant.

The discussions of these topics of spurious enthymemes are brief, in some cases only a few lines long, and amount to no more than exemplifications. There follows a brief account of 'refutation'. The section is at the very end of Book II, and perhaps it is possible to read into the cursoriness of treatment an impatience with the subject and desire to have done with it.

Much of the earlier parts of the *Rhetoric*, however, has some relevance for us, and we shall find occasion to refer back to it.

In assessing validity of an enthymematic argument which, from the point of view of a theory of deduction, may be incompletely stated we run into a difficulty of principle. In reading between the lines to a precise formulation of premisses and conclusion we are all too likely to beg the question ourselves. Aristotle believes that Rhetoric is an autonomous study and that he can sort out a set of peculiarly rhetorical argument-types which are purveyors at once of persuasive force and logical validity;

but he does not explain how there can also be, in the first place, sound but unrhetorical arguments which presumably do not persuade and, in the second, spurious enthymemes which persuade but are invalid. Without oversimplifying a complex problem it can be said that, by choosing which rule to apply, we may often support or reject an argument more or less at our discretion. For example, arguments condemned as fallacious under number [4] of our topics of spurious enthymemes, the use of a 'sign' as certain evidence, could often alternatively be regarded as rhetorically valid under the dispensation which permits us to refrain from stating arguments in full. This is the kind of criticism we have made earlier of some modern treatments of fallacy, and it is now clear that the defect has a long history; though not yet clear that it cannot be cured.

The existence of this lacuna in the theory is consistent with what is conjectured about the historical relation of the *Rhetoric* to the other works. If we assume that, about the time he wrote the *Rhetoric*, Aristotle was also engaged in revising the corpus of other logical works we can make parallel sense of the order of arrangement, the retention of the *Topics* as well as the *Prior Analytics* (for Dialectic and Rhetoric are related applications of the latter work's pure theory) and the insertion of the cross-references designed to encourage us to see the works in relation. Knowing that Aristotle's interest were changing towards the natural sciences and that he may have been preoccupied with teaching we can, perhaps, also explain the cursory nature of what are probably the *Rhetoric*'s last-written paragraphs and its author's last-written words on a logical subject.

BEGGING THE QUESTION, AND MANY QUESTIONS

Having dealt with the general features of Aristotle's writings we are now ready to take up features of his discussion of particular Fallacies.

The Fallacies of Begging the Question and Many Questions depend in conception, more than any of the other kinds, on the context of contentious argument. If we are to be literal, a question cannot be 'begged' without someone to do the begging, that is, to ask to have a premiss granted to him containing the substance of what is in dispute; and the Fallacy of Many Questions can occur only when there is actually a questioner, who asks two or more questions disguised as one. If we find putative examples of these Fallacies occurring outside the context of question-and-answer debate this can only be because some alternative account is being given side by side with this simple one.

Many Questions, as it happens, does not appear outside the *Sophistical Refutations*, and the account of it there is fairly straightforward; but Begging the Question, besides being described in the *Sophistical Refutations* and *Topics* is treated also in the *Prior Analytics*, where Aristotle was in the process of developing a context-free Logic of pure form. In the latter case the word 'beg' is completely inappropriate. The word 'assume', as we noticed, occurs sometimes as an alternative to 'beg', though it is hardly more frequent in the later work than the earlier. 'Beg', however, tends to dominate the account in the *Topics* (162b 34: the sentence which precedes the quoted passage is interpolated but the passage itself seems to be original):

> People appear to beg their original question in five ways: the first and most obvious being if any one begs the actual point requiring to be shown: this is easily detected when put in so many words; but it is more apt to escape detection in the case of different terms, or a term and an expression, that mean the same thing. A second way occurs whenever any one begs universally something which he has to demonstrate in a particular case: ...

The other ways are: begging particular cases of what should be demonstrated universally; begging a conjunctive conclusion piecemeal; and begging one of a pair of statements that 'necessarily involve one another', that is, are interdeducible.

Most of these make very little sense transported to the *Prior Analytics*, where the only question supposed to be relevant is that of formal validity of inferences. If two statements 'necessarily involve one another', it is possible to deduce one from the other, and that is all there is to be said; the deduction of a particular case from a universal that contains it is explicitly allowed for, and so on. When we turn to the *Prior Analytics* we find, however, that he does make an attempt to describe Begging the Question in a new and context-free way, but ends up making some quite different and puzzling remarks about it (65a 10):

If then it is uncertain whether A belongs to C, and also whether A belongs to B, and if one should assume that A does belong to B, it is not yet clear whether he begs the original question, but it is evident that he is not demonstrating: for what is as uncertain as the question to be answered cannot be a principle of a demonstration. If however, B is so related to C that they are identical, or if they are plainly convertible, or one belongs to the other, the original question is begged. For one might equally well prove that A belongs to B through those terms if they are convertible.

Formally, he seems to say, a begging of the question is a degenerate syllogism in which, because of the trivial satisfaction of one of the premisses, the other premiss and the conclusion are each as good or as bad as the other, so that argument from one to the other is nugatory; and that this is so is shown, among other things, by the fact that the premiss will, in this case, be as uncertain as the conclusion. However, this account will work only for the case in which B and C are 'identical'. If they are merely 'convertible' (that is, co-extensive though not strictly the same in meaning, like 'man' and 'featherless biped') it is difficult to see why fault should be found with the syllogism which results: say, 'If all men are featherless bipeds and all Greeks are men, all Greeks are featherless bipeds'. The same applies even more strongly to the case in which the only relation between B and C is that 'one belongs to the other'; that is, that the required premiss connecting B and C is *true*. The translator of the Oxford Aristotle here adds the footnote 'As genus to species' to explain 'belongs', but does not further explain his footnote. Sir David Ross, referring to this passage, says in part (*Aristotle*, p. 38):

> And syllogism is distinguished from *petitio principii* in this, that while in the former both premisses together imply the conclusion, in the latter one premiss alone does so.

This is a bold attempt to help Aristotle find a purely formal criterion of question-begging but the contrast with what Aristotle actually says is alone enough to reveal its inadequacy. Unless we import the idea that the premiss is being invoked to prove the conclusion in circumstances in which it is no more acceptable than the conclusion itself, all we are left with is a plainly valid argument.

Another point in which Aristotle departs, deliberately or not, from the attempt to build a purely formal theory concerns the relative certainty of premisses and conclusion. In what sense are premisses ever more certain than the conclusion they entail? In the only sense of 'certain' relevant to context-free logic a conclusion is always at least as certain as the (conjunction of the) premisses that lead to it. Aristotle, however, elsewhere says (67b 3; *Posterior Analytics* 86a 21) that argument proceeds from the more certain to the less certain and, moreover, that it is sometimes possible for a man to know the premisses of an argument to be true without knowing the conclusion to be true. This assumption is necessary if we are to admit that anyone ever learns anything by reasoning; but it seems to reveal a confusion of logical certainty with a more flesh-and-blood, epistemological variety.

Aristotle is a rewarding writer to study in so far as apparently loose and discrepant remarks of this kind can usually be provided with an explanation if one is prepared to dig for it. Here it is relevant to quote his actual definition of Begging the Question in the *Prior Analytics* (64b 33):

> ... since we get to know some things naturally through themselves, and other things by means of something else (the first principles through themselves, what is subordinate to them through something else), whenever a man tries to prove *what is not self-evident* by means of itself, then he begs the original question.

(My italics.) The clue is his concern with *how we come to know* things. Aristotle's theory of knowledge is alien to us; but at this point it touches his logic.[1] We come to know some things immediately, others mediately by inference. Propositions have a peck-order; and what can be known immediately cannot, strictly speaking, be known mediately, or vice versa. The truly certain things are those known immediately, the others being so much the less certain from having to be arrived at by inference. The one and only proper function of inference is to derive mediate truths from immediate. ('Immediate inference', which is an invention of later logicians, is not true inference at all.)

This account is sketchy and oversimplified, but it will do for

[1] See *Posterior Analytics* 71b 19, and generally in Book I, chapters 2 and 3.

our purpose. It is the account Aristotle was developing in his later work, as he became interested in Science. We can put it to work to make sense of what he says about Begging the Question.

If it is uncertain whether all Bs are As, and equally uncertain whether all Cs are As, we cannot use one to prove the other, since premisses must always be *more* certain – more immediately known – than their conclusions. If Bs and Cs are the same things whether because the concepts are identical *or* merely because the terms are convertible, 'All Cs are As' seems to be inferable from 'All Bs are As', but also vice versa; but there cannot be genuine inferences *both* ways, or there could be argument in a circle. Hence the apparent inference is really fallacious.

Again, consider the possibility that one of B and C can be predicated of the other. Perhaps Aristotle means *not* that all Cs are Bs, which would license his inference, but that all Bs are Cs: thus the proposed inference is 'If all Bs are As and all Bs are Cs, all Cs are As'. This is not formally valid at all; but it commits yet a further sin in having a premiss 'All Bs are As' which, given the other premiss, can be deduced from the conclusion and must therefore (by his account) be less certain than the conclusion. This is a case of question-begging since the premiss could not strictly be known except by way of deduction from the conclusion.

Hence the Fallacy of Begging the Question survives in an epistemological interpretation rather than a dialectical one. The new interpretation can be given a certain fanciful analogy, however, to the old: there is still, as it were, a question-and-answer debate, but now we are corporately questioners and Nature is the answerer. Nature properly grants us only self-evident truths as premisses: the rest we must infer for ourselves. Begging a Question is asking Nature to grant us, without argument, a proposition that is not self-evident. In the end Aristotle objects less to faults of the reasoning itself than he does to the fact that the reasoner goes the wrong way about getting knowledge.

The apriorism of Aristotle's Science is, however, not an essential element in his account, for other theories of knowledge can give rise to a similar tendency to regard as 'begged' any proposition which does not have appropriate specified credentials. Empiricist versions of the story come later.

NON-CAUSE AS CAUSE

One way of refuting a thesis is to deduce an obvious falsehood from it; and, in practice, it will often be the case that the falsehood is deduced not from the thesis alone, but only granted certain other assumptions. However, there is a logical trap inherent in this procedure. If a questioner seeking to demolish thesis T extracts from the answerer the concession of some other statement U before deducing false statement X, he has not necessarily proved the falsity of T at all, but only the falsity of one or other of T and U. Since the answerer has conceded both of these he has been exposed in *a* falsehood, but not necessarily in the falsity of T. He may say 'That is not the reason: X is false not because T is, but because U is. All you have done is show me I should not have admitted U'.

This is a simplified version of the kind of fallacy Aristotle had in mind when he wrote about 'treating a non-cause as a cause' in the *Sophistical Refutations*. It has no relation whatever to what is generally referred to as the Fallacy of False Cause in modern books. The word 'cause' is not here being used in its natural scientific sense at all, but in a purely logical sense. We shall see in a moment that Aristotle himself is partly responsible for the confusion. However, let us first look at the form of argument a little more closely.

The reasoning involved is sometimes referred to as *reductio ad impossibile*; though, in the description given, *reductio ad falsum* would be a more precise name, since we have presented the derived conclusion X as a merely false, not impossible one. If X were impossible one might deduce that the premises T and U were together also impossible. Aristotle writes as if this is what he had in mind, but the weaker case of derivation of a statement which was merely false would probably be as common in practice. If impossibility is demanded it will usually be obtained only by having the supplementary assumption U inconsistent with the thesis T, and 'non-cause as cause' can be pleaded by putting the blame on the conjunction of T and U instead of on T alone.

Since the workings of the Fallacy can be analysed in propositional logic it can be regarded as a formal one, like Consequent.

This being so, it is at first sight not surprising that it is described also in the *Prior Analytics*. On deeper reflection it can be seen that there is an inconsistency, for, if the Logic of the *Prior Analytics* is intended for use purely in deriving subordinate truths from first principles there should be no place for *reductio*, whether *ad impossibile* or *ad falsum*. First principles are supposed to be, at the very least, true, and from true statements no false ones follow. A *reductio* argument may be used in working out the system to derive one kind of valid syllogistic schema from another, but not in a 'demonstrative' proof.

That Aristotle is uneasy about its inclusion can be seen in his terminology. The *Sophistical Refutations* refers quite literally to the Fallacy as the one 'about the non-cause as cause', ὁ παρὰ τὸ μὴ αἴτιον ὡς αἴτιον (167b 21), but makes it clear that a logical interpretation is intended by, later in the book, referring to it merely as 'insertion of irrelevant matter'. Irrelevant matter can be inserted in an ordinary argument without prejudice to the conclusion, but it is methodologically dangerous to permit it in a *reductio*.[1] In the *Prior Analytics* the word 'cause' is completely avoided and the Fallacy is referred to by a prepositional phrase, 'the not because of that', τὸ μὴ παρὰ τοῦτο (65a 33 ff). Aristotle has become self-conscious about the use of the noun 'cause' in connection with the relation between premisses and conclusion, because he wants to distinguish demonstrative inference, where he thinks there is such a relation (see *Posterior Analytics* 71b 22), from dialectical and contentious reasoning, where there may not be. This is, in fact, a step in the development of the scientific concept of a 'cause': the discussion in Book II of the *Posterior Analytics*, and in the *Physics* and *Metaphysics*, is well known.

We are now ready for the account in the *Rhetoric*: it is very brief, and can be quoted in full (1401b 30).

> Another line [i.e. topic] consists in representing as causes things which are not causes, on the ground that they happened along with or before the event in question. They assume that, because *B* happens *after A*, it happens *because of A*. Politicians are especially

---

[1] However, some later writers objected generally to insertion of irrelevant matter: cf. the Stoic Fallacy of Superfluous Premiss, below, p. 92.

fond of taking this line. Thus Demades said that the policy of Demosthenes was the cause of all the mischief, 'for after it the war occurred'.

This is, in effect, the account the modern books give. The kind of fallacious argument described in the earlier works is much too complicated to be recommended to orators and is not mentioned. Instead, the phrase 'non-cause as cause'—here τὸ ἀναίτιον ὡς αἴτιον — is given a scientific, non-logical sense which will avoid ambiguous connotation.

The term 'False Cause' is no older than Isaac Watts, who used it (*Logick*, p. 473) in 1725. It was taken up by nineteenth-century English translators; but no word meaning 'false' occurs in the original. No doubt it has played its part in consolidating the revised account in popular favour.

REFUTATIONS DEPENDENT ON LANGUAGE

How firmly is Aristotle wedded to the classification of fallacies into the two main groups of the *Sophistical Refutations*, those Dependent on Language and those Outside Language? The answer is that although he says several times that this is the best and most appropriate classification he also says much that compromises any attempt to make it a rigid one. Thus, all fallacies can be considered to fall under Misconception of Refutation (168a 19); Misconception of Refutation itself might be brought under Fallacies Dependent on Language (167a 85); and Equivocation and Amphiboly are related to Many Questions (175b 39). *Secundum Quid* is described in a way and with examples that make it very like Composition or Division; and Form of Expression is not unlike Solecism, which is not classed in either category, though it is clearly 'dependent on language' (see 169b 35). It must be added that the dichotomy is dropped, without replacement, in the *Prior Analytics* and *Rhetoric*.

What, precisely, does Aristotle mean by 'dependent on language'? Familarity with modern treatments of fallacy seduces us into regarding all divisions as loose and not very meaningful; but would it not be reasonable to suppose that Aristotle had something quite definite in mind? Unfortunately what he actually says is extremely puzzling and leads us to suppose that we do not have the full story (165b 25):

Those ways of producing the false appearance of an argument which depend on language are six in number: they are ambiguity, amphiboly, combination, division of words, accent, form of expression. *Of this we may assure ourselves both by induction and by syllogistic proof* based on this – and it may be on other assumptions as well – that this is the number of ways in which we might fail to mean the same thing by the same names or expressions.

(My italics.) How we are to construct such proofs remains a mystery. A little later (168a 23) he does divide the six types into two groups of three, those that depend on double-meaning – Ambiguity, Amphiboly, and Form of Expression – and those that arise only 'because the phrase in question or the term as altered is not the same as was intended'; but this is not enough to give us the proofs referred to. We shall deal later with some attempted reconstructions.

In the meantime something needs to be said in support of the plausibility, in historical context, of the main twofold division.

A natural tendency would be to regard the refutations dependent on language as concerned purely with words, rather than with the concepts or thoughts behind them and which they express. Aristotle rules this out explicitly by refusing to countenance the distinction, and points out (170b 12) that it would make nonsense of the concept of equivocation, whose analysis involves *both* the words *and* their (multiple) meanings. What does distinguish the refutations dependent on language is that they all arise from the fact that language is an imperfect instrument for the expression of our thoughts: the others could, in theory, arise even in a perfect language. Admittedly, this principle of division can hardly be sustained if the fallacies are all given their modern interpretations; but if we run through them we shall see that Aristotle's examples, setting aside one or two special points, are all consistent with it.

Equivocation and Amphiboly gives us little trouble. Aristotle starts (165b 31) with the *Euthydemus* example

> Those learn who know: for it is those who know their letters who learn the letters dictated to them. For to 'learn' is ambiguous; it signifies both 'to understand' by the use of knowledge, and also 'to acquire knowledge'.

F

and goes on to

> 'The same man is both seated and standing and he is both sick and in health: for it is he who stood up who is standing, and he who is recovering who is in health: but it is the seated man who stood up, and the sick man who was recovering. For 'The sick man does so and so', or 'has so and so done to him' is not single in meaning: sometimes it means 'the man who is sick or is seated now', sometimes 'the man who was sick formerly' . . .

For Amphiboly we have (166a 7):

> 'I wish that you the enemy may capture'.

and

> 'There must be sight of what one sees: one sees the pillar: therefore the pillar has sight' . . . Also, 'Speaking of the silent is possible': for . . . it may mean that the speaker is silent or that the things of which he speaks are so.

This is also from the *Euthydemus*. All the examples clearly depend on double-meanings, whether in a word or in syntactical construction, that are contingent features of language and can be regarded as imperfections of it. Aristotle says (166a 15) that there are three kinds: (1) when either the word or the phrase has strictly more than one meaning: as when 'the dog' can refer either to an animal or to the dog-star, Sirius; (2) when we associate two meanings by custom; and (3) when the double-meaning arises in a word taken in combination with other words. We might have some reservations about whether (2) is really different from (1), or about whether (2) can ever really be avoided in a live natural language but, truistically, language *ought* to be built on the principle 'One word, one meaning'.

Accent and Form of Expression also create no difficulty. What Aristotle says about Accent was described above. Language should relate written and spoken forms one-for-one. In the case of Form of Expression Aristotle gives two classes of examples. First, Greek, like many other languages, has different inflected forms for masculine, feminine, and neuter words, but tends to have irregularities so that one may be misled about gender. Secondly, a word like 'flourishing' can appear to be 'active' when it is really 'passive': that is, transitive when it is really intransitive. It is not unreasonable to demand of a perfect language that it should be 'regular' in matters of gender and conjugation.

The only cases that cause us any difficulty, then, are Composition and Division. The examples of these fall into two classes, and it is possible that Aristotle intends Composition and Division to be different kinds of fallacy rather than, as usually understood, converse forms of a single kind. The first example of Composition is (166a 22):

> ... 'A man can walk while sitting, and can write while not writing'. For the meaning is not the same if one divides the words and if one combines them in saying that 'it is possible to walk-while-sitting' ... if one does not combine them, it means that when he is not writing he has the power to write.

Another example (177b 10) contrasts 'I saw-a-man-being-beaten with my eyes' with 'I saw a man being-beaten-with-my-eyes'. Our first question must be why these are not simply examples of Amphiboly. Aristotle says (177a 40)

> This fallacy has also in it an element of amphiboly ... but it really depends upon combination. For the meaning that depends on the division of the words is not really a double meaning ...

He puts up for comparison the case of two words which differ slightly in pronunciation, one having a 'smooth breathing' (unaspirated initial vowel) and the other a 'rough breathing' (initial aspirate), and continues:

> In writing, indeed, a word is the same whenever it is written of the same letters and in the same manner – and even there people nowadays put marks at the side to show the pronunciation – but the spoken words are not the same. Accordingly an expression that depends on division is not an ambiguous one.

What distinguishes these cases from Amphiboly, then, is that although the double-meaning occurs in the written form, it is removed – or, at least, can be removed – when the words are spoken aloud. This is done, obviously, mainly by grouping words together or separating them by pauses. Marks of pronunciation could, in principle, remove the double-meaning in the written form too.

The other class of example may be subject to the same analysis but, at least, is not quite so clearly so (166a 33):

Upon division depend the propositions that 5 is 2 and 3, and even and odd, and that the greater is equal: for it is that amount and more besides.

When we say that 5 is 2 and 3 we cannot deduce that 5 is *both* 2 *and* 3. As we should say, 'and' does not here indicate conjunction, but numerical addition. No doubt it is better to realize that addition is a separate concept and use a separate word for it. Nevertheless it was not unreasonable for Aristotle to suggest that the distinction be made by pronouncing the sentence in different ways.

We noted that, in the *Rhetoric*, Aristotle seems to be intending to give an account of Composition and Division when he speaks (1401a 25) of the Fallacy of Whole and Part. If this is so, his interpretation has moved even more completely away from the original one than it has in the case of Non-Cause as Cause.

ACCIDENT AND CONSEQUENT

What Aristotle says about the Fallacy of Accident has seldom been clearly understood. This is partly his own fault, since he is unclear what he means by 'accident' and hovers between two interpretations. The Fallacy occurs whenever (166b 23)

> ... any attribute is claimed to belong in a like manner to a thing and to its accident. For since the same thing has many accidents there is no necessity that all the same attributes should belong to all of a thing's predicates and to their subject as well.

Two examples are immediately given, of which one is:

> ... 'If [Coriscus] be different from Socrates, and Socrates be a man, then,' they say, 'he has admitted that Coriscus is different from a man *because it so happens* that the person from whom he said that he (Coriscus) is different is a man '

The phrase 'because it so happens' (διὰ τὸ συμβεβηκέναι) is cognate with the name 'Accident' (παρὰ τὸ συμβεβηκὸς). But, if 'accident' means what it is usually taken to mean, it is very strange indeed that being a man should be regarded as an accident of Socrates, rather than as an essential, perhaps defining, property as in the tree of Porphyry. Moreover, many logic books would need to be rewritten if we were to ban every argument as fallacious that has 'Socrates is a man' as a premiss.[1]

Later examples of Accident are sketched in the form of questions, indicating currency in dialogue and debate (179a 32):

> All arguments such as the following depend upon Accident. 'Do you know what I am going to ask you?' 'Do you know the man who is approaching', or 'the man in the mask?' 'Is the statue your work of art?' . . . 'Is the product of a small number with a small number a small number?'

He explains:

> . . . in the case of a good thing, to be good is not the same as to be going to be the subject of a question; nor in the case of a man approaching, or wearing a mask, is 'to be approaching' the same thing as 'to be Coriscus', so that suppose I know Coriscus, but do not know the man who is approaching, it still isn't the case that I both know and do not know the same man; nor, again, if this is mine and is also a work of art, is it therefore my work of art, but my property or thing or something else. (The solution is after the same manner in the other cases as well.)

It is not going to be possible to analyse all these examples in the same way but something can be done to make Aristotle's thought-processes clearer. In the first place, we should notice that he has a persistent tendency, not merely here but elsewhere in his writings,[2] to use the word 'accident' for any property of a thing that is not *convertible* with it: 'man' is an accident of 'Socrates' because although Socrates is a man not every man is Socrates. The fault of the inference

> Coriscus is different from Socrates.
> Socrates is a man
> Therefore, Coriscus is different from a man.

is that the second premiss can *not* be taken as an identity, or regarded as equivalent to 'Every man is Socrates'. We cannot entirely exonerate Aristotle by regarding the English word

---

[1] It has sometimes been suggested in explanation that the Socrates referred to is not the well-known one but a younger, still beardless, member of Aristotle's classes.
[2] Particularly in the *Posterior Analytics*, 73a 20–74a 4; see discussion in Poste, pp. 112–15, 128–9.

'accident' as a mistranslation: he appears to think that the predicate of a convertible proposition necessarily gives the essence of its subject and that the predicate of a non-convertible one can only give an accidental property.

Elsewhere (168b 27, 169b 6) Aristotle makes the, to modern minds, bewildering statement that Consequent is a variety of Accident. (There is no hint whatever of any kind of relation, as in modern books, between Accident and *Secundum Quid*.) This is quite intelligible if examples of Consequent are cast in syllogistic form rather than propositional. We might see Aristotle here as groping towards the theory of syllogistic inference that later flowered in the *Prior Analytics*. Some commentators have openly identified Accident with syllogistic invalidity.[1]

With some strain we could accommodate the 'paternal dog' example to this analysis: the source of invalidity of the quasi-syllogism 'The dog is yours; the dog is a father; therefore, the father is yours' could be traced to the non-convertibility of 'The dog is a father', in virtue of the fact that not all fathers are this dog. The 'small number' example is not explained and any explanation must be quite conjectural.[2] The other examples, however, introduce different considerations. The identification of 'Coriscus' with 'the man approaching' is, in context, convertible and there seems to be more substance here in the complaint that the fault is in the accidental nature of the relation. The example has sometimes been dismissed as Equivocation with the remark that 'know' means 'be familiar with' in one premiss and 'recognize' in the other; but this will not do, since we could easily formulate a parallel example in which this was not so; say

Coriscus is known-by-me-to-be-musical.
The man in the mask is Coriscus.
Therefore, the man in the mask is known-by-me-to-be-musical.

[1] Poste, p. 158, note 14 to ch. 24. Compare Buridan (fourteenth century): 'The fallacy of accident is therefore a deception arising from the fact that the conclusion seems to follow syllogistically from the premisses and does not follow from them.' *Compendium Totius Logicae*, tract 7.

[2] Poste, pp. 73, 156–7, reads into it 'A four multiplied by a four is a large number; but a four multiplied by a four is a four; therefore a four is a large number', where 'In one premiss, four multiplied by four means the product of the factors, in the other, only the first-named factor'; but it is still not clear why this should be Accident.

The most obvious way of making this valid would be by strengthening the second premiss to 'I *know* that the man in the mask is Coriscus', whence the inference, though not any longer in traditional syllogistic form, might be sanctioned by some modern 'epistemic' logic. This partially vindicates Aristotle's diagnosis of the fault but also means that we can, perhaps, charge him with confusing 'epistemic' modalities with 'alethic' ones.[1] Discussion had better be left, at this point, to later chapters.

MISCONCEPTION OF REFUTATION

The Fallacy which has become known as *Ignoratio Elenchi* is described by Aristotle in the following passage (167a 20):

> Other fallacies occur because the terms 'proof' or 'refutation' have not been defined, and because something is left out in their definition. For to refute is to contradict one and the same attribute – not merely the name, but the reality – and a name that is not merely synonymous but the same name – and to confute it from the propositions granted, necessarily, without including in the reckoning the original point to be proved, in the same respect and relation and manner and time in which it was asserted. . . . Some people, however, omit some one of the said conditions and give a merely apparent refutation, . . .

There is an attempt here to set down complete conditions of valid 'refutation'; that is, of 'proof'. Some of these echo the discussions of other fallacies: thus, 'without including in the reckoning the original point to be proved', in effect, is aimed at excluding question-begging. Some of the others, however, particularly the stipulations about 'same respect and relation and manner and time', are not explicitly dealt with elsewhere; and it is to infractions of these that Aristotle addresses his examples. A thing can be 'double' another thing at one time, but not at another time, or in one respect, say breadth, but not in another, say length. We shall see later that these examples came to assume an entirely disproportionate importance in the history of Logic.

It strikes Aristotle, however, that, if his definition of 'refutation' is really complete, he can regard all fallacies as due to failure, in some respect or other, to conform to it; and he runs through

[1] von Wright's terms: see his *Essay in Modal Logic*.

the entire list to illustrate this (167a 17–169a 22). At this point it is clear that there is an overlap in the classification: Misconception of Refutation should have been confined to the cases not covered elsewhere.

The name 'Irrelevant Conclusion' that has been used in modern books as a replacement is not very apt: Aristotle's point is that people *misunderstand* what a refutation consists in. Aristotle himself, however, is not very clear about the difference between knowledge of how to define 'refutation' and ability to put the definition into effect.

'BABBLING'

The phenomenon that the Oxford translators eupeptically term 'babbling'[1] is of more philosophical importance than its position at the end of the list (173a 31) indicates. It will be recalled that this is one of the five kinds of fate the contentious arguer may plan for his opponent. To babble, which Aristotle seems to regard as one stage worse than solecism, is to get involved in a verbal iteration or regress through a repetitious definition. If 'double' is defined as 'double of half', then the word 'double' in the second phrase may be replaced by its *definiens*, giving 'double of half of half'. which in turn becomes 'double of half of half of half'; or if 'snub' means 'having a concave nose', a snub nose is a nose having a concave nose. The trouble, Aristotle says, is that relative terms like 'double' do not have meaning in abstraction by themselves, and should therefore not be defined alone. More adequately we might say that if an incomplete expression *is* defined alone it should be defined in terms of a similarly incomplete one, and not equated with one that has been surreptitiously completed.

Aristotle may be seen, as in several other cases, to be groping towards a logical point but finding it incompletely in a verbal or dialectical one. The babbling or iteration, if it occurs, is produced by the questioner who brings out the consequences of the faulty definition after visiting it on his victim.

---

[1] ἀδολεσχεῖν, to talk idly, prattle on. The standard Latin translation is *nugatio*, which is unnecessarily insulting.

CHAPTER 3

The Aristotelian Tradition

It would be stretching history to claim that Aristotle's work on fallacies was an instant and unqualified success. We surmise that it continued to be taught in the Lyceum. We know that a copy of it was 'rediscovered' in a cellar in Asia Minor in the first century B.C., and republished by Andronicus of Rhodes. Minor references to it in various authors assure us that it was generally available to be read: the educated Greek or Roman of Athens or Alexandria or, later, Rome or Constantinople could walk down to his local library and consult a copy if he chose. Yet the entire bulk of extant literature on the subject, from the time of Aristotle up to the eleventh century A.D., would hardly fill a small notebook. This is a period of fourteen hundred years, equal to the period from Boethius to the present. We know that some writings have been lost, but, even so, it is clear that the subject was not often the centre of academic attention.

We shall trace this history briefly and then move on to the much more important contributions of the Middle Ages.

Aristotle's work was not quite the only work on fallacies from his period: we must not forget the Megarians and Stoics. In particular, Eubulides taught at Megara, not far from Athens, in Aristotle's time and is credited with having invented examples resembling those of Aristotle. Our surviving accounts of the activities of the Megarians and Stoics, however, are all from some hundreds of years later. Diogenes Laertius (second century A.D.) writes (*Lives of Eminent Philosophers*, II, § 108):

> To the school of Euclides belongs Eubulides of Miletus, the author of many dialectical arguments in an interrogatory form, namely,

The Liar, *The Disguised Electra*, *The Veiled Figure*, *The Sorites*, *The Horned One*, and *The Bald Head*. Of him it is said by one of the Comic poets: 'Eubulides the Eristic, who propounded his quibbles about horns and confounded the orators with falsely pretentious arguments, is gone with all the braggadocio of a Demosthenes.' Demosthenes was probably his pupil and thereby improved his faulty pronunciation of the letter R. Eubulides kept up a controversy with Aristotle and said much to discredit him.

Diogenes's gossipy style does not encourage faith in his historical accuracy and two pages later he ascribes the arguments known as the 'Veiled Figure' and 'Horned One' to Apollonius Cronus instead. However, it is interesting to learn of these examples and particularly that they are supposed to be 'dialectical arguments in interrogative form'. We surmise that in the time of Eubulides the Megarians taught and learnt their Philosophy in much the same way as the members of Plato's Academy. One of Eubulides's pupils, Alexinus, was so argumentative that they nicknamed him Elenxinus. The arguments Diogenes mentions are described in various places: 'Electra' and 'The Veiled Figure', for example, are in Lucian's play *Philosophies for Sale*, in which the Gods evaluate various philosophical positions by the novel method of having the philosophers themselves offered for sale as slaves. Chrysippus, representing the Stoics, is interrogated by a potential buyer (§ 22–3):

BUYER What do you mean by the Veiled Figure and the Electra?
CHRYSIPPUS The Electra is the famous Electra, the daughter of Agamemnon, who at once knew and did not know the same thing; for when Orestes stood beside her before the recognition she knew that Orestes was her brother, but did not know that this was Orestes. As to the Veiled Figure, you shall hear a very wonderful argument. Tell me, do you know your own father?
BUYER Yes.
CHRYSIPPUS But if I put a veiled figure before you and asked you if you know him, what will you say?
BUYER That I don't, of course.
CHRYSIPPUS But the veiled figure turns out to be your own father; so if you don't know him, you evidently don't know your own father.
BUYER Not so: I should unveil him and find out the truth!...

Lucian, perhaps deliberately, does not state the arguments very well, but we can fill in the details. Aristotle, as we noticed, has an example virtually the same as these two, under the heading of Accident. He resolves it as follows (179a 36):

> For it is evident in all these cases that there is no necessity for the attribute which is true of the thing's accident to be true of the thing as well ... nor in the case of a man approaching, or wearing a mask, is 'to be approaching' the same thing as 'to be Coriscus' so that suppose I know Coriscus but do not know the man who is approaching, it still isn't the case that I both know and do not know the same man; ...

Of the other arguments mentioned by Diogenes Laertius, 'The Liar' is the well-known demonstration that a man may lie and tell the truth at the same time, by asserting that he is lying; 'The Disguised' is probably another version of 'The Veiled Figure'; 'The Sorites', or heap, is the demonstration that there can be no such thing as a heap of sand since one grain does not make a heap and adding one grain is never enough to convert a non-heap into a heap; 'The Bald Head' is a similar proof that no one can have a head of hair; and the 'The Horned One' is, in one formulation, the argument 'What you have not lost you still have; you have not lost horns; therefore you still have horns', which is given special point by a double-meaning. These have been enumerated and described in more detail elsewhere.[1] We can see of these examples that they raise logical points which would serve as a focus for discussion but it is not obvious that they fit into any system.

Sextus Empiricus, who is the best of our sources on the Megarians and Sotics, tells us that there did exist a classification of fallacies. His account of the Stoic classification is hostile and possibly suspect; but it is all we have.[2]

> Now the Dialecticians [i.e. Stoics] assert that an argument is inconclusive owing to inconsistency or to deficiency or to its being propounded in a bad form or to redundancy. An example of inconsistency is when the premisses are not logically coherent with each

---

[1] See, for example, Kneale, p. 114. A much longer list is given in Gassendi, *Syntagma Philosophicum* (1658), Part I, Chapters 3 and 6. See also Aldrich, Appendix 5.

[2] *Outlines of Pyrrhonism*, II, § 146–51. Parallel account in *Against the Logicians*, II, § 429–34. See also Mates, *Stoic Logic*, pp. 82–3.

other and with the inference, as in the argument 'If it is day, it is light; but in fact wheat is being sold in the market; therefore Dion is walking.' And it is a case of redundancy when we find a premiss that is superfluous for the logic of the argument, as for instance 'If it is day it is light; but in fact it is day and Dion also is walking; therefore it is light.' And it is due to the bad form in which it is propounded when the form of the argument is not conclusive; for whereas the really syllogistic arguments are, they say, such as these: 'If it is day it is light; but in fact it is day; therefore it is light'; and 'If it is day, it is light; but it is not light; therefore it is not day,' – the inconclusive argument runs thus: 'If it is day, it is light; but in fact it is light; therefore it is day.'

... And the argument is faulty by deficiency, when it suffers from the omission of some factor needed for the deducing of the conclusion: thus, for instance, while we have, as they think, a valid argument in 'Wealth is either good or bad or indifferent; but it is neither bad nor indifferent; therefore it is good,' the following is faulty by way of deficiency: 'Wealth is either good or bad; but it is not bad; therefore it is good.'

Several comments should be made on this passage. In the first place, the four kinds of 'inconclusiveness' are regarded as involuntary lapses of reasoning rather than as tricks of a sophist. There is no suggestion of question-and-answer, and the category-headings themselves are inappropriate to this interpretation. We know independently that the later Stoics, particularly Philo and Chrysippus, had a fairly well-developed propositional logic that was formal in outlook,[1] and a corresponding change in their conception of fallacies seems natural. The word *sophisma* is not used in this connection and, at least by the time of Sextus, the word 'Dialectic' has dropped its old meaning and has become simply the standard word for Logic. Elsewhere Sextus sometimes uses examples which reveal that they have been adapted from sophistical dialogue: thus the example (*Outlines of Pyrrhonism*, II, §231)

> It is not true both that I have asked you a question first and that the stars are not even in number.
> But I have asked you a question first.
> Therefore the stars are even.

[1] See Kneale, ch. 3, and Mates, ch. 5.

makes sense only if the 'question' referred to in the second premiss is the one that has arisen from the putting of the first premiss in question form. Possibly the change in emphasis from the dialogue form of argument to a context-free form was made by the later Stoics.

Secondly, it should be noted that two of the 'inconclusive arguments' are in fact formally valid. The only thing wrong with the second kind, due to redundancy, is that there is a superfluous premiss; and in the case of the fourth we have, as we would normally reckon these matters, a valid argument but a false first premiss. It would be possible to put these anomalies down to muddleheadedness or a faulty text but we should do this only as a last resort. The concept of logical argument is a complex one and it is early to find reasons for refusing to regard arguments of the second and fourth kinds as perfectly valid: in the case of the second, the statement of an argument ordinarily carries a presumption, if seldom an explicit claim, that the stated premisses are unredundant, and, with reference to the fourth, it often happens that an argument is erected on an *ad hoc* 'tautology' that turns out, on examination, to be less than truly tautologous – a factor has been overlooked. Although we cannot exonerate Sextus from the blame of inadequate explanation we should not be too ready to condemn the classification.

Finally it should be noted, however, that the account suffers from the same fault as most other accounts of fallacies: its too-brief statement makes it a catalogue rather than a doctrine. We do not know what the Stoics may have made of these distinctions in their verbal tradition or in lost writings – Diogenes Laertius gives us an impressive catalogue of books on the subject by Chrysippus[1] – but whatever system there was has not been reproduced here.

Between the time of the later Stoics and the time of Diogenes and Sextus there is no logician of interest, but we might note in passing the rhetorical writings of Cicero, which were subsequently very influential. One of these, the *Topics*, is explicitly intended to be a commentary on Aristotle's *Topics* and was later accepted

---

[1] *Lives of Eminent Philosophers*, VII, § 189–202. For what is preserved of Chrysippus's writings on Fallacies, see Von Arnim, *Stoicorum Veterum Fragmenta*, vol. 2, pp. 89–94.

as part of the Aristotelian logical tradition. Cicero wrote it, however, on a shipboard holiday off Calabria and, having no actual copy of Aristotle with him at the time, was forced to rely on memory. His memory was not good: the result has very little connection with anything that Aristotle ever wrote, and what connection it has is with the *Rhetoric* rather than the *Topics*. Cicero wrote a great deal on Rhetoric but nothing on fallacies, and his influence has tended to cut fallacies out of the subsequent rhetorical tradition. At the same time he picked up logical points from Aristotle's *Rhetoric*, and the historical accident that his work was studied in the Middle Ages at times when Aristotle's *Rhetoric* was unknown has given his work traditional importance. Aristotle, for example (*Rhetoric* 1355b 36), had referred to certain 'modes of persuasion not belonging strictly to the art of rhetoric', namely, those that depend on the evidence of witnesses or evidence gained through tortures or especially guaranteed by oath or by written declaration; and arguments based on these were called 'extrinsic' by Cicero and contrasted with more direct or 'intrinsic' arguments.[1] These are the nearest recognition in Greek and Roman writers of the modern *ad hominem*, *ad verecundiam*, and so on. Cicero has lawsuits in mind: the advocate may use these as valid aids to his case but should treat them as of less account than arguments that tell directly on the desired conclusion.

Now let us consider what Sextus Empiricus has to say about fallacies on his own account. He is the most famous of the 'Pyrrhonian' sceptics; so-named after Pyrrho, an obscure contemporary of Aristotle who, according to Diogenes Laertius (*Lives*-IX, ch. 11, § 61–2), suspended judgement so completely even on the existence of things around him that he needed regularly to be saved from carts, precipices, and dogs by a party of friends who followed him about for the purpose. Whether or not Sextus is as sceptical in practice, he is a nihilist in theory and sets out not merely to disprove all particular doctrines, philosophical or any other, but also to demonstrate the impossibility of truth, reason and proof. To the obvious criticism that a proof of the non-existence of proof is self-defeating he replies that some purgatives

---

[1] Cicero, *Topics*, ch. 2, § 8; 19, § 72. The words *extrinsecus* and *intrinsecus* are actually adverbs.

'after driving the fluids out of the bodies expel themselves as well' and says (*Against the Logicians*, II, §481):

> And again, just as it is not impossible for the man who has ascended to a high place by a ladder to overturn the ladder with his foot after his ascent, so also it is not unlikely that the Sceptic after he has arrived at the demonstration of his thesis by means of the argument proving the non-existence of proof, as it were by a step-ladder, should then abolish this very argument.

This must be the source of the idea in Wittgenstein's famous passage.[1] Sextus's long and detailed attack on the concept of proof incorporates many arguments usually credited to later writers including, in several variant forms, the complaint that every valid proof must be a case of Begging the Question. For example (*Against the Logicians*, II, §357):

> By what means, then, can we establish that the apparent thing is really such as it appears? Either, certainly, by means of a non-evident fact or by means of an apparent one. But to do so by means of a non-evident fact is absurd; for the non-evident is so far from being able to reveal anything that, on the contrary, it is itself in need of something to establish it. And to do so by means of an apparent fact is much more absurd; for it is itself the thing in question, and nothing that is in question is capable of confirming itself.

Fallacies are clearly grist to his mill and he is adept at detecting them in the writings of others; but *doctrines* of Fallacy are a different matter and must be thrown away just as one throws away the ladder one has climbed up on. Sextus's writings on this subject are, I believe, quite unique and contain insights of a fundamental character. These are of the same kind as, but ultimately more incisive than, his strictures against Proof.

His basic point concerns the practical application of a doctrine of fallacy. (The following quotations are all from *Outlines of Pyrrhonism*, II, §236–59.)

> As regards all the sophisms which dialectic seems peculiarly able to expose, their exposure is useless; whereas in all cases where the exposure is useful, it is not the dialectician who will expose them but the experts in each particular art who grasp the connexion of the facts.

---

[1] Wittgenstein, *Tractatus*, 6.54; see Weiler, 'Fritz Mauthner's Critique of Language', p. 86.

We might regard this as Sextus's way of distinguishing between 'logical' and 'empirical' knowledge. He considers the following syllogism:

> In diseases, at the stages of abatement, a varied diet and wine are to be approved:
> In every type of disease an abatement inevitably occurs before the first third day;
> Therefore, it is necessary to take a varied diet and wine before the first third day.

The doctrine of a three-day fever-cycle was current in contemporary medicine. The term 'abatement' is equivocal, referring in the first premiss to the period of convalescence and in the second to the temporary upturn every third day; but – and this is the point – the logician has no means of spotting the trouble, whereas a doctor, with specialized knowledge, will find it immediately *because he already knows that the conclusion is false*. The function of the study of arguments is generally supposed to be to enable us to reason from premisses to conclusion with confidence; but ambiguity can be detected only *a posteriori*, by seeing that there are arguments of valid form with true premisses and false conclusions.

> For just as we refuse our assent to the truth of the tricks performed by jugglers and know that they are deluding us, even if we do not know how they do it, so likewise we refuse to believe arguments which, though seemingly plausible, are false, even when we do not know how they are fallacious.

> ... Thus, suppose there were a road leading up to a chasm, we do not push ourselves into the chasm just because there is a road leading to it but we avoid the road because of the chasm; so, in the same way, if there should be an argument which leads us to a confessedly absurd conclusion, we shall not assent to the absurdity just because of the argument but avoid the argument because of the absurdity.

It seems to be Sextus's view that no word is absolutely ambiguous or absolutely unambiguous. Distinctions are made when, and only when, something hangs on them.

> And in the ordinary affairs of life we see already how people – yes, and even the slave-boys – distinguish ambiguities when they think

such distinction is of use. Certainly, if a master who had servants named alike were to bid a boy called, say, 'Manes' (supposing this to be the name common to the servants) to be summoned, the slave-boy will ask 'Which one?' And if a man who had several different wines were to say to his boy 'Pour me out a draught of wine,' then too the boy will ask 'Which one?' Thus it is the experience of what is useful in each affair that brings about the distinguishing of ambiguities.

If this is true, the role of a discussion of ambiguity in the study of Logic must always be a strictly limited one; for if a validly-drawn conclusion seems to us false, we may explain this by 'discovering' an ambiguity in one of the terms. On the other hand, if ambiguity is not detectable apart from its connection with argument, the practical use of argument is also severely circumscribed; for, if we cannot guarantee any term against ambiguity, we cannot allow any argument to influence us to accept a conclusion that we do not already accept on independent grounds.

This charge, moreover, is not so easily answered as is the one that every argument is question-begging; for the latter does not depend on the added complication of ambiguous terms. There have been many discussions of, and answers to, the latter; but, to my knowledge, no one has ever taken up Sextus's criticisms of the concept of fallacy. We shall do so in due course.

An approximate contemporary of Diogenes and Sextus in the late second century A.D. was the Aristotelian commentator Alexander of Aphrodisias. He is credited with being the first to use the word 'logic' in its modern sense, and his logical inventions, include the *pons asinorum* or 'bridge of asses', the diagram which embodies Aristotle's principles of argument-finding. Here history has really treated us badly because, although several of his commentaries on Aristotle's works survive, a commentary on the *Sophistical Refutations* which was known and quoted both in his own time and in the Middle Ages and of which, in consequence, we have enough glimpses to make us impatient to read it in full, has been lost. It is the only known case of a commentary on the *Sophistical Refutations* before the eleventh century. Our scanty knowledge of its contents has been pieced together from references in later Byzantine and medieval commentaries, foremost among which is the *Treatise on the Major Fallacies* of Peter of Spain.[1]

Alexander's principal contribution to our subject was the doctrine of τὸ διττόν, 'the double', which came to be known in Latin as *multiplex*: the terms are used to mean 'double meaning' or 'multiple meaning' but clearly also have overtones of deceit, something like the word 'duplicity' in English. The ramifications of the concept of *multiplex* enable Alexander to give a unified account of Aristotle's Refutations Dependent on Language. We can reasonably suppose that Alexander came to it as a result of puzzling over Aristotle's cryptic remarks, quoted above, about the demonstrable completeness of the enumerated six kinds (*Soph. Ref.* 165b 23. The Byzantine sources refer their discussion to this passage):

> Of this we may assure ourselves both by induction, and by syllogistic proof based on this – and it may be on other assumptions as well – that this is the number of ways in which we might fail to mean the same things by the same names or expressions.

Does Aristotle have some principle of exhaustive enumeration in mind? We have seen that he tells us very little, even in passages where he compares and contrasts the members of the list (168a 23, 169a 22). Alexander, on the other hand, gives us a neat, textbookish and utterly clear division of them.

There are three kinds of *multiplex*, actual, potential, and

---

[1] Peter of Spain, *Tractatus maiorum fallaciarum*, München Staatsbibliothek Clm. 14458, and four other manuscripts, has never been printed. There are briefer references in Peter's *Summulae Logicales*, ch. 7. The Byzantine sources are *Alexandri quod fertur in Aristotelis Sophisticos Elenchos Commentarium*, vol. 2, part 3 of *Commentaria in Aristotelem Graeca* (ed. Wallies), probably by Michael Ephesius (eleventh century); and a paraphrase of the *Sophistical Refutations* (ed. M. Hayduck), in the same collection, vol. 23, part 4, probably by Sophonias about the year 1300. It is temporally possible that Peter should have read the earlier of these, but it is distinct from the (lost) commentary of Alexander that he mainly refers to, though it was confused with it for many centuries. On its authorship see the introduction to the above edition. The *multiplex* doctrine also appears in the short *Fallacies Dependent on Language* ascribed uncertainly to Alexander's contemporary Galen: *Works*, vol. 14, pp. 582–98.

It has been suggested to me by Nicholas Rescher that there may yet exist an Arabic translation of Alexander's commentary. Such a translation was made from Syriac in the tenth century by Yaḥyā ibn 'Adī: see Rescher, *Development of Arabic Logic*, p. 131.

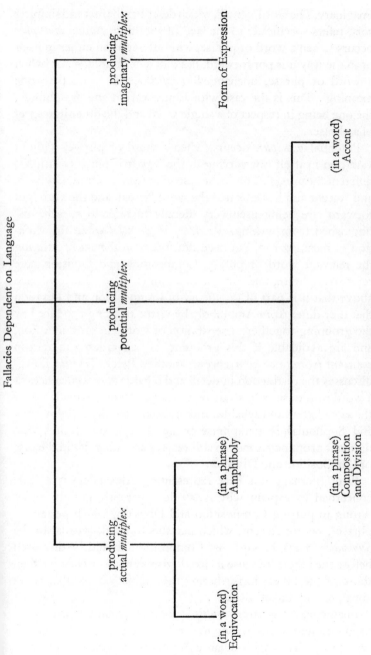

Alexander's tree of Fallacies Dependent on Language

imaginary. The word 'actual', which describes the first and simplest sort, refers specifically to the 'act' of speaking. Actual *multiplex* occurs 'when a word or phrase, without variation either in itself or in the way it is put forward, has different meanings', or 'when a word or phrase, unchanged *simpliciter*, has more than one meaning'. This is the case with Equivocation and Amphiboly, the one being in respect of a single word and the other in respect of a phrase.

Potential *multiplex* occurs 'when a word or phrase, without variation in itself but varying in the way it is put forward, has different meanings'. This is the case with Composition, Division, and Accent and it seems that the word 'actual' and the verb 'put forward' are being used very literally to refer to speech, distinguished from writing. *Multiplex* is *potential* when the variation of meaning can, but need not, occur in the act of uttering the relevant word or phrase. Composition and Division here refer to a phrase, and Accent to a single word. We have seen above that it is part of the theory of Composition and Division that they differ from Amphiboly by virtue of being produced by the grouping-together or separation of words in pronunciation, and are avoidable if this grouping or separation can remain constant from one occurrence to another. Peter (*Treatise*, f. 6va) discusses the distinction in detail and regards two occurrences of a word or phrase as 'materially' the same if they are made up 'of the same letters and syllables and divisions' (cf. *Soph. Ref.* 177b 5 and Sophonias(?) paraphrase p. 49, l. 23), but thinks that different pronunciations can still resolve an ambiguity in the case of Composition and Division.

Peter (perhaps not here representing Alexander) is a little concerned to explain why Aristotle apparently got the order wrong in putting Composition and Division, which pertain to phrases, before Accent, which pertains to single words. In the *Summulae* (7.24), he says that Composition should be discussed before the others 'because its paralogisms are more effective than those of the others in furthering the sophist's aims, that is, an apparent and unreal wisdom'.

Imaginary or apparent *multiplex* occurs 'when a word with a fixed meaning seems, owing to some likeness, to have a different one' (7.34). The Aristotelian examples of deponent verbs with active meaning but passive conjugations, and nouns of feminine declension denoting males, are given. It is not explained why this kind of *multiplex* occurs only in single words and not in phrases, but this will obviously be so so long as it is explained in terms of misleading word-endings.

This completes the classification. However, if Alexander succeeded in erecting on it a proof of exhaustivity, this did not communicate itself to Peter of Spain, who is even more cryptic than Aristotle. He says (*Summulae* 7.05):

> Aristotle shows, by induction and by syllogism, why there are six fallacies *in dictione*. The proof by induction is as follows: Equivocation arises in one of those six modes; Amphibology; and so on for the others; therefore, every fallacy *in dictione* arises in one of those six modes. The proof by syllogism is as follows: every deception due to our meaning something not the same by the same names or phrases arises in one of those six modes; but every fallacy *in dictione* arises due to our meaning something not the same by the same names or phrases; therefore every fallacy *in dictione* arises in one of those six modes.

With callous irrelevance he adds that the syllogism is in the first mood of the first figure. We may surmise that Alexander, at least, would have been acute enough to see this argument as a good example of Begging the Question. Peter pushes the deductive argument back a little further by giving equally question-begging syllogisms to justify the major and minor premisses. If we turn for enlightenment to the Byzantine commentary we find a similar syllogism attributed to Alexander's contemporary, Galen; and this may be verified,[1] but here the chase stops.

We find the elements of the *multiplex* doctrine in Aristotle if we dig. He occasionally uses the term of which '*multiplex*' is a

---

[1] Actually, if the work of Galen referred to is the *Fallacies Dependent on Language* mentioned above, this does not give a syllogism in full but says that Aristotle's account is quite obscure and 'the phrase "this is the number of ways in which we might fail to mean the same thing by the same names or expressions" is more like the conclusion of a syllogism than like a syllogism; but such brevity of speech is customary in this philosopher . . .': *Works*, vol. 14, pp. 584–5. What are clearly Galen's own thoughts on Fallacies are given in the curious *Diagnosis and Treatment of Faults of Every Mind*, which is a precursor of some twentieth-century Philosophy in likening errors to tumours and the logician to a medical practitioner: *Works*, vol. 5, pp. 58–103.

translation, and it is translated 'double meaning' or 'ambiguity'. Thus he says (168a 24)

> For of the fallacies that consist in language, some depend upon a *double meaning*, e.g. ambiguity of words and of phrases, and the fallacy of like verbal forms (for we habitually speak of everything as though it were a particular substance) – while fallacies of combination and division and accent arise because the phrase in question or the term as altered is not the same as was intended.

(The italics indicate the occurrence of τὸ διττόν.)
Elsewhere he says (177b 8)

> Accordingly an expression that depends upon division is not *an ambiguous one*.

(Italics for διττόν.) But it is clear that it is 'potentially' ambiguous, though he does not say so directly. In discussing Form of Expression he says (178a 24)

> ... really the meaning is not alike, though it *appears* to be so because of the expression.

For 'appears' we have φαίνεται, cognate with φαντασία, the word used in reference to 'imaginary' *multiplex* in the Byzantine works. None of this detracts from the doctrine's originality, but it does indicate a very thorough reading of Aristotle by someone determined to find system in its more obscure passages. That person was presumably Alexander, whose commentaries on others of Aristotle's works[1] are so painstaking and perceptive.

We can now trace the interesting history of the loss and rediscovery of the doctrine of fallacies in western Europe. This is the best example in history of how a whole area of learning can become dormant and, in spite of a felt need for it, require the efforts of new and original scholars to bring it back to consciousness.

There is nothing surprising, of course, about the loss of manuscripts. The slow erosion of old documents goes on by war, flood, and simple dereliction even in the twentieth century, and the period between the sixth and twelfth centuries was not exactly a peaceful one for Europe. It is easy, however, to exaggerate the loss and its causal role. Manuscripts disappear, in part, because people are not interested enough to save them; and it is

[1] Vol. 2 of *Commentaria in Aristotelem Graeca*.

not fair to blame marauding Goths or Moslems without first reflecting that these people, too, had their scholars and their intellectual priorities. Theodoric, the Roman emperor at the time of Boethius, was a civilized Goth who had the opportunity to help preserve Roman learning if he had wanted to; and the Arabs in fact did possess translations of virtually all Aristotle's works.[1] The Eastern Roman Empire at Constantinople, moreover, had a continuous and relatively peaceful existence unaccompanied by book-burning until it fell to the Turks in 1453; but for centuries it took little interest in the manuscripts it undoubtedly had.

The scholars of the West in the twelfth century found their new awareness of Logic exciting even before they discovered the lost works of Aristotle which were later to be taken as their leading texts and the proper cause of their excitement. In the interregnum between their first awareness of their need and the demonstration that Aristotle's *Sophistical Refutations* filled it they relied on a half-spurious Aristotelian doctrine put together out of fragments; and when, finally, the manuscripts were unearthed

[1] I have Nicholas Rescher's permission to quote verbatim a note he wrote for me on Arab contributions:

'The *Sophistical Refutations* (SR) was first translated into Arabic around A.D. 830 by 'Abd al-Masīḥ ibn Nā'imah, a Jacobite Christian. A later translation (from the Syriac version of Theophilus of Edessa) by Yaḥyā ibn 'Adī (893–974) – who also translated the commentary on this treatise by Alexander of Aphrodisias – became the standard Arabic version.

'The major Arabic Aristotelians all dealt with this treatise. The great commentary on SR by Avicenna survives and was published in Cairo in 1958. Averroes also commented on this text in his standard, triplicate manner (short, middle, and great commentaries), and a medieval Latin translation of the middle commentary, made from a Hebrew intermediary, is unique among Arabic SR materials to be accessible in a European tongue. The later Arabic logicians, who worked outside the context of specifically Aristotelian texts, did not much concern themselves with fallacies, and they dropped out of view in the standard handbooks produced after the early thirteenth century.

'The principal – perhaps sole – explicit deployment of the conception of fallacy in Arabic extra-logical writings occurs in the setting of the hostile commentaries or "refutations" (a *genre* patterned on a Greek model, the refutations by Alexander of Aphrodisias of various treatises by Galen). Thus, for example, in Averroes's refutation of al-Ghazzālī's *Refutation of the Philosophers* (*Tahāfut al-Tahāfut*, translated by S. van den Bergh) one finds Ghazzālī being charged (vol. 1, p. 3) with committing the fallacy of many questions, "one of the well-known seven sophisms".'

and translated the genuine and spurious Aristotles lived for a time beside each other. The spurious account is important for its contribution to that most characteristic of all medieval inventions, the doctrine of Supposition.

For the origin of the spurious doctrine of fallacies we must go back to before the loss of the manuscripts. Boethius, the Roman of the sixth century who was imprisoned and executed by the emperor Theodoric after many years of service to him, wrote extensive commentaries on works of Aristotle, Cicero, Porphyry and others besides writing logical works of his own. History has been kind to these writings.[1] From his time until the twelfth century Aristotle's *Topics, Sophistical Refutations, Prior* and *Posterior Analytics* and *Rhetoric* were, for practical purposes, unavailable in western Europe; but certain works, consisting of Aristotle's *Categories* and *Interpretation*, Porphyry's *Introduction*, Boethius's commentaries on these last three works and on Cicero's *Topics*, together with Boethius's own other logical writings, became standard 'classics' for logicians. The twelfth century subsequently called this corpus of works the 'Old Logic', to contrast it with the rediscovered lost works. It contained no sustained discussion of fallacies, and such references as there are to individual fallacies of Aristotle's list do not indicate any kind of systematic connection. However, in two passages, both by Boethius, there are discussions which, at first sight, *look* like accounts of part of Aristotle's original list. Their connections have been traced by De Rijk.[2]

At one point in the *Sophistical Refutations* Aristotle had had a brief and quoteworthy definition of *refutation* which we have already noticed (167a 23; see also echoes at 180a 22, 181a 3):

> For to refute is to contradict one and the same attribute – not merely the name, but the reality – and a name that is not merely synonymous but the same name – and to confute it from the propositions granted, necessarily, without including in the reckoning the original point to be proved, in the same respect and relation and manner and time in which it was asserted.

[1] It has been suggested that Cassiodorus used his influence with the Emperor Theodoric to help preserve Boethius's writings: see Isaac, *Le Peri Hermeneias en Occident*, p. 30.
[2] *Logica Modernorum*, vol. 1. The present chapter leans heavily on this book.

This definition we find actually quoted by Alexander of Aphrodisias and a later Alexandrine writer, Asclepius, in their respective commentaries on Aristotle's *Metaphysics*. It is an interesting definition because it encapsulates a virtually complete doctrine of fallacy. A refutation can fail because

    i. the reality is not contradicted, but only the name, or
   ii. the proof contains only a 'synonymous' word, or
  iii. the premisses of the refutation are not granted, or
  iv. are not necessary (but only accidental), or
   v. the original point to be proved is among the premisses, or
  vi. the refutation does not refute in the same respect, or
 vii. relation, or
viii. manner, or
  ix. time.

Some unknown Greek writer has recognized this, at least so far as categories i, ii, and vi–ix are concerned; for these are listed by Boethius, and also (except ii) independently by Ammonius Hermiae about a century earlier. The same or a similar list is given in the two medieval Byzantine works, as a bonus subclassification of Misconception of Refutation. Boethius tells us elsewhere that he found many marginal notes scribbled on his copy of Aristotle and spent a lot of time and trouble deciphering them, so perhaps this is his source. The passage on which he is commenting (*Interpretation* 17a 34) is concerned with affirmation and denial, and the burden of the comment is that opposition between pairs of statements may sometimes be apparent rather than real. Consequently all his examples are of apparently contradictory pairs.

    i. *Equivocation.* 'Cato killed himself at Utica' and 'Cato did not kill himself at Utica' may fail to be mutually contradictory, since there are two historical Catos one of whom did, and one of whom did not, kill himself at Utica. This is unlike any of Aristotle's examples of equivocation, since it depends on ambiguous reference of a proper name rather than ambiguous connotation of a predicate. However, Boethius gives as an alternative example 'Cato is strong' and 'Cato is not strong' where the word 'strong' (*fortis*) can refer either to physical or mental prowess.

ii. *Univocation.* 'Man walks' and 'Man does not walk' can fail to be contradictory, since 'man' can refer either to a particular man or to the species Man. (The Latin *homo ambulat* is more ambiguous than the English because of the lack of definite or indefinite articles.) This is a strangely named Fallacy and one that does not either contrast cleanly with the previous one or square well with the appropriate clause of the passage in Aristotle. In his alternative treatment in the *Introduction to Categorical Syllogisms* Boethius explicitly restricts his examples to singular statements as distinct from generalizations and consequently omits this heading. However, Univocation was to be important later in connection with the concept of Supposition.

iii. *Different part.* 'The eye is white' and 'The eye is not white' may both be true even if the same eye is referred to, since one proposition may refer to the eyeball and the other to the pupil.

iv. *Different relatum.* 'Ten is double' and 'Ten is not double' fail to contradict one another if the term 'double' has different *relata* in the two cases. In the *Introduction to Categorical Syllogisms* he gives the example 'Socrates is on the right-hand side' and 'Socrates is not on the right-hand side'.

v. *Different time.* 'Socrates is sitting down' and 'Socrates is not sitting down' may both be true at different times.

vi. *Different modality.* 'The kitten can see' (*catalus videt*) and 'The kitten cannot see', though both refer to the same kitten at the same time, may both be true since one may be a statement of potentiality, the other of actuality. In the *Introduction to Categorical Syllogisms* he gives the example 'The egg is an animal' and 'The egg is not an animal'.

Boethius's examples, except in the case of Univocation, are all singular statements. In the *Introduction to Categorical Syllogisms* he confines himself to singular statements explicitly and omits Univocation from the list. He says directly that Aristotle followed up these matters very thoroughly in the *Sophistical Refutations*: this is not, of course, true, since the Boethian list is relevant only to the original headings of Equivocation and Amphiboly, and Aristotle has no subheadings or examples like most of these and in some directions is even more cursory than Boethius. The 'old' logicians, however, were unable to check the reference and before the twelfth century they do not seem to have taken very much notice of Boethius's list at all.

Peter Abelard, undoubtedly the biggest figure of his time, can hardly be regarded either as an 'old' logician or as a 'new'. His life (1079–1142) is astride the period of rediscovery. His early logical training must have been exclusively in the 'old' works and he never attained more than a superficial knowledge of the new ones; yet his writings display a new spirit which puts him ahead of his time, and there is no doubt of his influence on his successors. His time was transitional in other ways: Logic schools in places like Paris were becoming more institutionalized, and were to grow within the next half-century into universities. Abelard himself taught at the cathedral school of Notre-Dame and at his own school on the Mont-Sainte-Geneviève.

Both in his *Logica Ingredientibus* (the work is named by the first word of its text) and in his *Dialectica* Abelard reproduces and discusses the Boethian sixfold classification of fallacies. The two works were probably both first written early in his career,[1] in Paris in the years leading up to the fateful incident of Héloise and her uncle Fulbert (1118). In the *Dialectica* he generally invents new examples: for Different Modality, for example, he has 'The peasant is a bishop', where he must have had in mind some friend or enemy who had been a peasant and became a bishop. For Different Time he gives in both books the same example, with a verb which is ambiguously present or past tense – *Socrates legit* can mean either 'Socrates is reading' or 'Socrates read' – changing the force of the ambiguity, which would have been there without this complication. However, he also has other contributions to make: he is critical of some of Boethius's examples and wants to raise analytical questions about them. When Aristotle spoke of the conditions of contradiction between pairs of affirmative and negative propositions did he stipulate that materially identical *words* must be used, or was he referring to identity of *meaning* independently of words? The first suggestion

[1] See the respective introductions to the editions cited in the bibliography.

is neither quite sufficient, since words may be equivocal, nor quite necessary, since (*a*) synonyms will do, and (*b*) sometimes words are context-dependent as when *A* says 'You are angry' and *B* replies 'I am not angry'. The second suggestion, on the other hand, raises all the difficulties about how we are to specify in each particular case what meaning a word has, and is close to being a question-begging one.

Univocation is even more thought-provoking. When we say 'Socrates is a man', does the word 'man' stand for an individual man or for the species Man? Socrates is not the species Man, and yet the fact that Socrates is a man consists in Socrates's having just those properties that characterize the species. Abelard's early interest in the problem of universals, inherited from his nominalist teacher Roscelin, is stirred. Roscelin had said that these terms were just 'noises' (*flatus vocis*). Abelard is clear that he must distinguish several kinds of use of these terms and speaks of different kinds of 'translation':[1] a word such as 'man' which has an ordinary concrete meaning in reference to individual men may be used in various transferred senses, thus:

[*a*] for the species Man;
[*b*] for the word itself, as in '"Man" is monosyllabic' or '"Man" is a word';
[*c*] metaphorically, as when (to change the example) the word *auriga*, which normally means 'charioteer', is used to refer to the captain of a ship; and
[*d*] for an entirely different but analogous thing, as when the word *canis* is used not for 'dog' but for 'the dog-star', Sirius.

An awakening interest in the study of grammar contributes to Abelard's thinking. Both grammatical and logical rules are well illustrated by examples of breaches, and the study of ambiguities of the Boethius type assumes an academic role in helping students make distinctions. At first, in Abelard, it is hardly more than that, and there is even some feeling that it is not really *fallacies*, either in the modern sense or in Aristotle's, that are

---

[1] See De Rijk, vol. 1, pp. 51–6. The term is used in the sense of 'metaphor', as one of the ten *tropes*, in the traditional *Rhetorica ad Herennium*, which in the Middle Ages was attributed to Cicero.

being studied. In this connection Abelard remarks (*Logica Inredientibus*, p. 400; quoted by De Rijk, vol. 1, p. 59):

> I remember once seeing and carefully studying a little book by Aristotle entitled *Sophistical Refutations*; but when I searched through the various kinds of sophism for Univocation I found nothing written about it. Consequently I have often wondered why Boethius said Aristotle recorded those six kinds there.

Had he really read the *Sophistical Refutations*? If so, he had failed to pick out the key passage defining *elenchus* and to identify 'univocation' with 'synonymy'. Others soon made the connection and smirked over Abelard's oversight.[1] This is, however, the first medieval reference to the *Sophistical Refutations* and indicates that there was probably a translation in limited circulation before the first one of which we have definite knowledge, that of James of Venice about 1128.[2] Abelard did not know it very well when he wrote his larger logical works. In the latter part of his life copies were certainly circulating freely. De Rijk has noted with interest that among the lost works of Abelard is mentioned a *Book of Fantasies*.[3] Abelard in one place in his *Dialectica* (III, 448 3-4) refers the reader to it for the fuller discussion of the question 'whether there is a beginning of time'. We do not know what else it contained, but it is interesting to speculate. The word *fantasia* was sometimes used as a synonym of *fallacia*.

---

[1] De Rijk, vol. 1, p. 60.
[2] There is a tradition that Boethius translated the *Sophistical Refutations* into Latin. Since the sixteenth century it has been generally assumed that the medieval 'vulgate' text was that of Boethius: this text is reproduced and attributed to Boethius in Migne, *Patrologia Latina*, vol. 64, cols. 1007-40. The attribution was based on flimsy evidence. Textual criticism has shown that the translation of the *Posterior Analyticus* in the same volume (cols. 711-62), also attributed to Boethius, is not by him but probably by James of Venice; see Minio-Paluello, 'Iacobus Veneticus Grecus, Canonist and Translator of Aristotle'. The translation of the *Sophistical Refutation* is by someone other than the translator of the *Posterior Analytics*, but we do not know who this is. There are several other twelfth-century or earlier translations and at least one of them, Milan, Biblioteca Ambrosiana I 195 Inf., could be by Boethius. Abelard's occasional quotations from the *Sophistical Refutations* in his later work appear to be from a lost translation which is yet another candidate. There is, however, no firm evidence that Boethius made any translation at all.
[3] *Liber Fantasiarum*; see De Rijk, vol. 1, pp. 109-12.

We now come to a uniquely original author of the twelfth century, Adam of Balsham, whose *Art of Discourse* was written in 1132.[1] Not much is known about him personally, but he was born at Balsham, near Cambridge, of a French family that had come over with William the Conqueror, and taught for many years in Paris at the *Parvipontanus* school, so named from its situation near the 'little bridge' over the Seine. The *Art of Discourse* is a deliberate attempt to break with tradition and produce a forward-looking textbook – not of 'Logic' but of 'Discourse', Logic's practical application. He deals with *questions* as well as *statements*; and for the latter he avoids the traditional term 'proposition', presumably for the same reason as many modern logicians, in favour, in his case, of *enuntiatio*. There is a long section on sophisms.

Adam's style reflects his preoccupation with avoiding the traditional and trite. He shares Aristotle's taste for using interrogative words as nouns but, in a Latin without articles, this leads to curiously convoluted formations. The four sections of the *Art* are entitled (as literally as I can translate them):

1. About what.
2. How, in what words, about what.
3. What about it.
4. How, in what words, what about it.

The exegete needs to do the sort of job on these that Europeans sometimes need to do on Chinese. It transpires that 'about what' (*de quo*) and the 'what' (*quid*) in 'what about it' are grammatical terms referring to two parts of a sentence; and these, in fact, became standard terms in medieval Logic. With a little mind-twisting we get the explication

1. The Subject.
2. Verbal form of the Subject.
3. The Predicate.
4. Verbal form of the Predicate.

The same kind of ingenuity is needed in reading Adam's treatment of sophisms, which is in the second section of the work.

[1] Minio-Paluello, *Adam Balsamiensis Parvipontani Ars Disserendi*. See also Minio-Paluello, 'The "Ars Disserendi" of Adam of Balsham "Parvipontanus"'.

The fourth section – if it was ever written – has not come down to us and we do not know whether there was or would have been a complementary account of sophisms in it.

Aristotle is nowhere mentioned in a logical connection but there are many indications of his influence. The enterprise of writing an 'art of discourse' itself puts one in mind of the *Topics* and *Sophistical Refutations*. It can also be demonstrated that Adam had been reading (pseudo-) Augustine.[1]

The extant manuscripts of the *Art of Discourse* give us two different versions of it, one later than the other and probably not by Adam. John of Salisbury, who had studied under Adam in Paris, complained in his *Metalogicon* (Book 1, Ch. 3) about logicians who draw all their examples from the technical language of Logic itself. This charge could be sheeted against Adam in the first version of the work but whoever wrote the second version[2] has replaced many of the examples with less pedantic ones, besides adding clarificatory words and sentences. Past a certain point in the text we possess the second version only.

The most interesting feature of Adam's treatment of sophisms is his survey of the ways in which words combined into phrases may suffer from ambiguity. A *single-term* designation, says Adam, derives its meaning from one of three sources, common use, technical use, or etymology. For the last of these he says 'arguments', but it transpires that 'etymology' is what he means. The appropriate classification of single-term ambiguities is according to the sources of the respective meanings: both from common use, one from common use and one technical, and so on in all possible combinations.

Complex, or multiple-term, designations, however, may have many quite different kinds of ambiguity. To start with, even when the ambiguity is in only one term of the designation – and an ambiguity in a single term will sometimes be found to occur *only* when that term is used in combination! – there is a distinction to be made between the case in which the root meaning of a word is the same but it is ambiguous as to grammatical case, number, mood or tense; the case in which a root meaning is

---

[1] See Minio-Paluello, 'The "Ars Disserendi" ', p. 136.
[2] Perhaps a certain Alexander Neckham; see Minio-Paluello's introduction to the text, p. xxii.

modified by context; and the case in which a word is simply vague or inexplicit. The beginnings of this arborification are to be found in Aristotle's subdivision of Equivocation and Amphiboly (*Soph. Ref.* 166a 14), but Adam is much more detailed. The last-mentioned category needs to be surveyed for different kinds, and Adam distinguishes eleven, including several which approximately categorize either Boethian or Aristotelian examples: one of them is called *Secundum Quid*, one is described like Accident, and another reproduces Abelard's actuality-possibility distinction. When the ambiguity depends on a combination of words and not on any one singly, there are many more distinctions to make: Adam lists twelve cases and gives them technical names. Here are a few of them: I transliterate rather than translate their titles.

1. *Conjunction.* 'This and that are not true' is ambiguous if this is true and that isn't. The example has a family likeness to some examples of Many Questions.

2. *Disjunction.* 'That it is day or night is always the case' could mean 'It is always day or it is always night'. It is remarkable that in these two headings *conjunction* and *disjunction* occur in their modern senses, in reference to 'and' and 'or'. De Rijk (vol. 1, pp. 72–3) regards them as suggestive alternatively of Composition and Division; but the examples do not correspond.

3–5. *Abjunction, transversion,* and *conversion.* These refer to kinds of alternative parsings of sentences; whether a word goes with an adjoining phrase or stands separately, whether a qualifier qualifies one word or another, whether a demonstrative word refers to one word or another.

7. *Traduction.* (that is, 'translation' as between languages). A source of ambiguity is the use of a construction appropriate in a foreign language but inappropriate in the one in which it is framed.

8. *Intellection.* The examples given are 'one cannot hear in Greece what is said in Egypt' and 'To know what is said is to know that it is said'. There are two feasible interpretations and complete comprehension is needed to enable us to choose between them.

The others (Perversion, Defection, Innexion, Connexion, Internexion) are of less interest.

An important section added in the second version describes various kinds of what it calls 'Transumption', which is analogous to (and obviously derived from) Translation in Abelard.

Adam's work was not as influential as it deserved to be: the schools – even his own after him – preferred pure Aristotle, or Aristotle alloyed only with Boethius. We should not be surprised at this. The labour of coming to terms with Greek (and Roman and Arabic) learning was to be enough in itself to keep the academic world busy for the next hundred years, and Adam's personal enterprise would have seemed brash to all but, possibly, his immediate associates and pupils.

We have reached, in fact, Aristotle's golden age: learning was to become identified, for several generations of scholars, with the task of reading and commenting on Aristotle. The Logic that was to grow out of this activity, moreover, was to form the backbone of the educational system. A university system, already implicit in the schools of the time of Abelard and Adam, became explicit around the turn of the century. A Faculty of Arts is supposed to be devoted to the 'seven liberal arts': the *trivium* of Grammar, Logic and Rhetoric and the *quadrivium* of Arithmetic, Geometry, Music and Astronomy. In practice the young (perhaps twelve-year-old) student's first need was to perfect his Latin, and Grammar was learnt incidentally – in a 'grammar school'. Literature, in the sense in which this is a special study, was not pursued: Humanities were not to be confused with Arts, which had a relatively serious purpose. The works to which the student turned were those of the 'old' and 'new' Logics, which were to be his basic training for whatever else – Theology, Law, Medicine – he might study later. The Renaissance later affected to despise the sterility and logic-chopping of the 'Schoolmen' but was in reality standing on their shoulders. In any case, this attitude developed only in the fifteenth century when Logic was in decline and the protagonists of the Humanities were hitting back.

It is possible, from surviving university statutes,[1] to get a

---

[1] See Isaac, *Le Peri Hermeneias en occident*, ch. 3; and Paetow, 'The Arts Course at Medieval Universities'.

fairly clear idea of the syllabus of studies in the thirteenth century. Details varied, but the course, of from four to seven years, was generally built round Logic and Aristotle's *Ethics*, *The Soul* and some physical and naturalistic treatises. Several hours each day were devoted to lectures by masters and 'tutorials' on them by bachelors. Approximately the same amount of time was devoted to the 'new' and 'old' Logics. The *Topics* and *Sophistical Refutations* were treated together. There is no mention in the time-tables of any writer more recent than Boethius, with the exception of Gilbert de la Porrée, whose *Six Principles* had become an honorary part of the 'old' Logic, The writings of more recent logicians were, no doubt, studied but they were not regarded as sufficiently venerable to be allowed to shape the syllabus.

The works that we are dealing with, then, are textbooks for Arts students. Some of the earliest of these have been edited by De Rijk (*Logica Modernorum*). James of Venice wrote a commentary on the *Sophistical Refutations*, but this is lost; and another short commentary from the twelfth century was burnt at Chartres in 1944. The earliest extant medieval commentary is from (probably) some time after 1150, the anonymous *Glosses on the Sophistical Refutations* (De Rijk, vol. 1, pp. 187–225) which continually quotes James of Venice and Alberic of Paris and was presumably pieced together by comparing previous commentaries of these respective writers. Abelard is also sometimes quoted. The much longer, also anonymous, *Summa of Sophistical Refutations* (De Rijk, vol. 1, pp. 257–458: a *Summa* is a complete or synoptic textbook) was written near the same date. De Rijk has edited several other anonymous works from later in the twelfth century. One of these, which he calls *Viennese Fallacies* (vol. 1, pp. 491–543), is remarkable in completely incorporating the Boethius classification into the Aristotelian scheme. Another is the *Parvipontanus Fallacies* (vol. 1, pp. 545–609) from the closing years of the century and illustrating, in its thorough Aristotelianism, Adam's lack of memorial.

We are told by Roger Bacon[1] that the first to 'read' the *Sophistical Refutations* at Oxford were St Edmund of Abingdon, who taught there from 1202 to 1209, and a certain Master Hughes. However, a number of discussions of fallacies are extant in works

[1] See van Steenberghen, *Aristotle in the West*, pp. 138 ff.

originating in England well before the turn of the century: among these is the impressive *Munich Dialectica*.[1] Robert Grosseteste (*c.* 1170–1253), Bishop of Lincoln and first Chancellor of the University of Oxford, is usually regarded primarily as a natural scientist, but wrote logical commentaries including one on the *Sophistical Refutations*. The largest of all commentaries is that of St Albert the Great (1193–1280), teacher in Paris of the yet more famous St Thomas Aquinas; who himself left among his lesser philosophical works an *opusculum*, or little treatise, on fallacies, addressed 'to certain noble Arts students' (1244–5 ?). Another famous master-pupil pair are William of Sherwood (1200/10–66/71) whose *Introduction to Logic* contains a chapter on fallacies which bears comparison with other whole treatises, and Peter of Spain, later Pope John XXI (or XX), whose *Summulae Logicales* is the most famous of all medieval texts, and is reported to have been in use as late as the nineteenth century: in Paris for centuries first-year students were known as *summulistae* in its honour. Peter's textbook leans heavily on that of William, but he also wrote the separate *Treatise on the Major Fallacies* which we have already noticed as giving us, in greater detail than any other extant work, the doctrine of *multiplex* of Alexander of Aphrodisias. Roger Bacon's *Sumule Dialectices* is a similar work of the period: these three give us a chance to assess the opinions of thirteenth-century writers on the place of the study of fallacies in relation to Logic as a whole. The list is by no means exhaustive, even of those whose work has been published; and there remain many unpublished and even virtually unread manuscripts.

The task of sifting examples and terminologies and tracing doctrinal influences within these works is an absorbing one, but it is for specialists in the period. The plain fact is that the works mentioned all deal with nearly the same material, and give tightly interlocking treatments of it. It will be enough to describe one of them in detail, with occasional adversion to this or that other; but which? Several fingers point in the direction of William of Sherwood. Roger Bacon, who seldom praises anyone, regarded him as the most original logician of his time. He was influential

---

[1] See De Rijk, *Logica Modernorum*, vol. 2, part 1, chapters 12, 13, and 15, and text of (particularly) the *Munich Dialectica* and *London Fallacies* in vol. 2, part 2, pp. 453–678.

not only on Peter of Spain and Roger Bacon but also on Albert the Great and Thomas Aquinas. Although it is no longer believed, as it has sometimes been, that he wrote the first consolidated logic textbook or that he invented the doctrine of Supposition, it is clear in both respects that his work was an important qualitative advance over what preceded it.

About William's life, again, we know very little.[1] He was probably born in Nottinghamshire between 1200 and 1210 and taught at Paris in the 1240s. The Arts Faculty at the University of Paris, which had had a troubled first half-century with a fight over the banning of Aristotle's *Metaphysics* and the burning of other works in 1210–15 and a strike over academic freedom that closed the University for three years following 1228, finally got to its feet and produced the Golden Age of Logic.[2] During the years 1240–7 William, Peter, Albert, Thomas, and Roger were all at some time in residence, and about 1245 they were possibly all present together. William was the senior and influential teacher and put the others, in varying degrees, in his debt. In retrospect it is strange that none of his works saw print before the twentieth century: no one of the other writers mentioned is in this position. The *Introduction to Logic* is his most significant work but there are several others certainly or probably by him. After his return from Paris he held several clerical positions in England, and died, in what circumstances we do not know, at some time in the period 1266–71.

Of the six books of the *Introduction to Logic* the content of five follows approximately that of the various books, excluding the *Posterior Analytics*, of Aristotle's *Organon*; thus:

Chapter 1. Statements (cf. *Categories*)
          2. Predicables (cf. *Interpretation*)
          3. Syllogism (cf. *Prior Analytics*)
          4. Places (cf. *Topics*)
          5. Properties of Terms
          6. Sophisms (cf. *Sophistical Refutations*)

The corresponding works of Peter and Roger deal with the same

---

[1] See the introduction by N. Kretzmann to his edition of the *Introduction to Logic*.
[2] See van Steenberghen for a history of this period.

material in slightly different arrangements.¹ Besides the loss of the *Posterior Analytics*, whose material is partly incorporated elsewhere, the content is poorer than that of Aristotle's *Organon* in just one important respect, in containing little reference to modality or modal syllogisms;² but this is counterbalanced by the incorporation of material from Boethius and the addition of the characteristic chapter 5, which deals with Supposition and allied concepts.

Chapter 3 is noteworthy for the first appearance of the famous *Barbara Celarent* verse embodying the theory of validity of syllogisms and of the reduction of the valid forms to those of the first figure.³

The closeness with which Aristotle is sometimes followed is startling to the modern mind. An example is the reproduction of the division of arguments into three types, demonstrative, dialectical and sophistical, in chapter 4, as in the *Topics*, in parallel with the inconsistent division into four types, including examination arguments, in chapter 6, as in the *Sophistical Refutations*. William, however, was himself puzzled over the conflict between the obviously dialectical character of the *Sophistical Refutations* and the overlaid claim that it can be seen as a supplement to the *Prior Analytics*. He says boldly (p. 132)⁴

> I maintain that the substance of disputation is nothing but syllogism. Considered as an entity, therefore, disputation and syllogism are one and the same thing. It is called 'syllogism', however, in virtue of the fact that a person can organize thought by means of it.

The examples in this chapter are nearly all, sometimes rather self-consciously, in the precise form of syllogisms as laid down in his chapter 3: a very few have compound subject or predicate.

The structure of the chapter on Sophisms is simple. The five possible aims of sophistical disputation are described, following

---

[1] The growth of this pattern can be traced through the twelfth-century *Tractatus Anagnini* and *Munich Dialectica* edited by De Rijk in *Logica Modernorum*, vol. 2, part 2, pp. 215–332 and 453–638.
[2] There is a detailed treatment of modal syllogisms in Buridan's *Compendium*, tract 5, of the succeeding century.
[3] See, however, De Rijk, vol. 2, part 1, pp. 401–3, on some precursors.
[4] In quoting from Kretzmann's translation I have frequently omitted his parenthetical references to the original Latin.

Aristotle: Refutation (*redargutio*), Falsity (*falsum*), Paradox (*inopinabile*), Babbling (*nugatio*), and Solecism (*soloecismus*). The refutations Dependent on Language (*in dictione*) and refutations Outside Language (*extra dictionem*) are dealt with separately, each section enumerating and describing the relevant sophisms as listed by Aristotle. The work then ends, without any of Aristotle's supplementary discussions. The same structure is found in the other works on our list, except for those in commentary form and the anonymous *Summa*, which is more elaborate.

In the case of the Sophisms Dependent on Language, all the works attempt to give a structure to their list in a way which reflects knowledge of the theory of *multiplex* in the lost commentary of Alexander of Aphrodisias. Most of them actually use the word *multiplex* or *multiplicitas*: the exception is William. Since a sophism, however, is an argument that *seems* valid but *is not*, it is considered appropriate in the case of each variety to specify the 'cause of the sameness' and the 'cause of the non-existence' (cf. *Soph. Ref.* 169a 22–b 17), and William gives what is, in effect, the classification by *multiplex* in these terms. Thus (p. 135)

> Sameness in the case of a linguistic whole is either in the substance of the discourse alone or in the substance of the discourse together with its use. (I speak of the use of discourse since in its use it is pronounced.) This [last] sameness is either of a word alone or of an expression. If it is sameness of a word and there is diversity in the corresponding reality it is called Equivocation; if it is sameness of an expression it is called Amphibology.
>
> Sameness of discourse with respect to its substance alone is either sameness of an expression and called Composition or Division, or it is sameness of a word and called Accent.
>
> On the other hand, sameness in the case of a grammatical ending in so far as it is a source of deception is in a word only and is called the Figure of a Word. (I am speaking of these samenesses [only] in so far as there is diversity on the part of the corresponding reality.)

This passage may be compared with the *multiplex* tree of Alexander: 'sameness' (*apparentia*) does very nearly the same job as '*multiplex*'. By the 'substance' of a word William means the word itself considered as a grammatical item. 'Words' and 'expressions' are to be understood as written entities, and the 'use' (*actus*) of a word is its use in speech. He uses systematically ambiguous terminology to avoid too precise an interpretation of this kind but what he says is not really intelligible on any other.

William reproduces Aristotle's threefold subdivision of Equivocation and Amphiboly (*Soph. Ref.* 166a 14; cf. Adam of Balsham, p. 26) but brings it forward to the beginning of his section in each case, so that he may exemplify them separately. The main examples, in the case of Equivocation, are (pp. 135–8):

1. A word properly signifies more than one thing on its own: 'Every dog runs – one of the heavenly constellations is a dog; therefore one of the heavenly constellations runs.'

2. A word signifies one thing properly and another 'transumptively': 'Whatever runs has feet, the Seine runs; therefore the Seine has feet.' Here the word 'runs' is 'transumptive' in the second premiss.

3. A word signifies different things as a result of its connections with something else: 'Whoever was being cured is healthy, the sufferer was being cured; therefore the sufferer is healthy.' The word 'sufferer' refers to someone who *was* suffering when it is put in a past-tense statement and to someone who *is* suffering when it is put in a present-tense statement.

Examples of Amphiboly (*amphibologia*) are nearly untranslatable, depending as they do on Latin idiom. The first, recalling Horace Greeley's comments about the relative newsworthiness of 'Dog bites man' and 'Man bites dog', is (p. 139)

Quemcunque verum est panem comedere, panis comedit illum. Sed canem verum est panem comedere. Ergo panis comedit illum.

which goes into bad English as

If someone bread did eat, bread ate him; but the dog bread did eat: therefore bread ate him.

This resembles the King Pyrrhus example, which is first found in the *Summa*.[1] If we turn to parallel accounts by other writers we find the same or closely similar examples used time after time, and influences may be traced by comparing occurrences.[2]

[1] De Rijk, vol. 1, p. 289. However, Aristotle has a very similar one: *Soph. Ref.* 166a 7.
[2] See De Rijk's indexes in vol. 1 and in vol. 2, part 2.

In discussing his examples William deals with issues arising out of his theory of meaning. Thus in reference to examples of type (3) of Equivocation he voices the objection that

> .. a word is prior to any expression, and therefore retains the essence it had before becoming an ingredient in the expression.

The answer to this objection is that some words signify concepts which are systematically 'participated in by what is more than one thing with respect to earlier and later'; like the word 'sufferer' in the example. The direction of William's thought on sophisms can be clearly traced in these subsidiary discussions, which regularly raise for him questions in the Philosophy of Language.

William's examples of Composition and Division also turn on time-distinctions: in the case of

> Whatever is possible will be true; it is possible for a white thing to be black: therefore it will be true that a white thing will be black.

the minor premiss is ambiguous since we may take 'a white thing' and 'to be black' together or separately; that is (taking 'Composition' in the literal verbal sense suggested earlier), we may pronounce it

It is possible (*pause*) for-a-white-thing-to-be-black.
– or –
It is possible-for-a-white-thing (*pause*) to be black.

Peter of Spain (*Summulae* 7.26–7.30), an inveterate arborifier whose classification proliferates branches and twigs, says there are two varieties of Composition and two of Division: the distinction depends on whether there is question of alternative bracketings of words or merely a question of presence or absence of a particular bracketing.

We have already dealt in some detail with the Fallacy of Accent and the style of the examples of it that are concocted to suit Latin idiom. William gives several examples, of which two are virtually the same as those of Peter of Spain given above in chapter 1.[1]

---

[1] Buridan, in his *Compendium* (tract 7), tells us that there are four ways accent can vary; first, in respect of continuity and interruption; second, in having syllables lengthened or shortened; third, in the importation or omission of aspirates; and fourth, in respect of modulation of the voice as acute, grave and circumflex.

Figure of Speech (*figura dictionis*, pp. 146–7) is referred entirely to grammatical terminations. Latin abounds in 'deponent' verbs, whose passive-voice conjugation conceals an active sense and we might be tempted to reason: *Sortes operatur: ergo patitur*, 'Socrates is working', (passive verb in Latin) 'therefore he is inactive'. Then follows William's curious discussion of the 'raw meat' example, 'What you bought yesterday you ate today; but yesterday you bought something raw, etc.' The word *crudum* means 'something raw' but William is worried by the fact that this term refers ambiguously to the 'what' of the raw thing and to its 'how' and distinguishes, in terminology which need not concern us, between two sorts of meaning a term may have. The point of the subsumption of this case under Figure of Speech seems to be that *crudum* is an adjective (denoting a 'how') doing duty for a noun (denoting a 'what') and that there should in strict logic – as in Latin grammar there is not – be distinct grammatical forms for adjectives and derived nouns.

In another case, 'What you had and do not have you have lost; you had ten and you do not have ten (suppose you have lost one); etc.' the word 'ten' (*decem*) ambiguously refers to a 'what' and a 'how much'. Roger Bacon would use the same explanation in the case 'Socrates is a man, Plato is a man; therefore Socrates is Plato'; the first premiss gives you a 'how' of Socrates and the conclusion says of him instead that he is a 'something' (*hoc aliquid*).

Peter of Spain (*Summulae* 7.34–8) finds that there are three kinds of Figure of Speech, of which the third is exemplified by 'Man is a species, therefore some man is a species'. The rationale of this is even harder to follow but there must have been a strong feeling that the ambiguity of the word 'man' – as individual, and as species – was of a kind that should really have been indicated by a variation of grammatical ending.

Turning to Sophisms Outside Language (*extra dictionem*) we find that William gives mainly traditional examples of Accident (p. 150): 'You know Coriscus; it is Coriscus who is approaching; therefore you know who is approaching'. The 'cause of the sameness' since this is a sophism *extra dictionem*, is the (non-verbal) sameness of Coriscus and the man approaching. *Secundum Quid* is also traditional. In the case of Ignorance Regarding

Refutation (*ignorantia elenchi*; p. 155), where some of the accounts interpolate a complete Boethian classification, William quotes Aristotle's definition of refutation (he here uses the Aristotelian word *elenchus* whereas in classifying kinds of refutation in his introduction he had used the Latinism *redargutio*) and then says that only the four specifications 'in the same respect, in the same relation, in the same way and at the same time' are relevant to the subclassification of sophisms under this heading. For the third kind he gives the example 'Socrates is naturally pious but he is not absolutely pious; therefore he is both pious and non-pious' and remarks that his examples are all also cases of *Secundum Quid* though he thinks this misses the main point which distinguishes them.

Begging the Question can occur in five ways, as in the *Topics* (162b 34): (1) when 'in order to prove something, that very thing is assumed', as in 'A man is running, therefore a man is running'; (2) when to prove a particular we assume a universal under which it falls; (3) when to prove a universal we assume, one after the other, *all* the particulars falling under it; (4) when to prove a conjunction we separately assume the conjuncts; and (5) 'when one has to prove something and assumes something else that necessarily affects it, as if one were to prove that the side is incommensurate with the diagonal and assumed that the diagonal is commensurate with the side'. However, the old trouble about question-begging comes up:

> But a doubt arises, because every one of those inferences is necessary, and in them there is a dialectical ground (*locus*, topic).

William's way out is that

> the acceptability of an inference lies not merely in the necessity of the consequence but is inseparable from its producing belief regarding a doubtful matter.

We have already noticed that Peter of Spain distinguishes here between an 'inference' and a 'proof'.[1]

The treatment of Consequent needs no special comment. Non-Cause as Cause is given a correct traditional treatment in connection with reasoning *ad impossibile*; and the same applies to all the

[1] p. 33 above.

medieval treatments, in spite of the superficial strangeness of the word 'cause' in this setting. The discussion of Many Questions (the heading is *De fallaciis secundum plures interrogationes ut unam*, 'Fallacies of many questions as one') is preceded by the statement that

> a question and a proposition are one and the same thing here, except that it is a question when it is asked prior to a syllogism, and it is a proposition when it is ordered in a syllogism.

In spite of this unpromising conflation, the examples take an entirely appropriate dialectical turn; thus

> ... suppose that two things are pointed to, one of which is good, the other bad. 'Are these things good or not good?' If one takes the affirmative, one is necessarily refuted, for it follows that what is not good is good. But if one takes the negative, refutation seems to follow although it does not follow, for this does not follow: 'they are not good; therefore this one of them is not good'.

In modern propositional terms, William denies that 'Not both p and q' implies 'Not p and not q'. We shall see that the discussion of this example links up with what William has to say about the Obligation game.

William's discussion stops abruptly (as does his whole treatise) at the end of his treatment of Many Questions but some of the other writers go on to subsidiary subjects. Peter (*Summulae* 7.65–9), for example, paraphrases Aristotle's reduction of various sophisms to Misconception of Refutation. The contemporaries of William and Peter seem to have had no inclination to discuss Solecism, Babbling, and the other forms of refutation.

What did these writers learn from studying fallacies? – Two very important things; neither of which we can discover by reading the treatments of fallacies themselves, since neither reacted to any extent on these treatments. William's chapter on Properties of Terms, however, contains many references to one or another of (particularly) the Sophisms Dependent on Language; and there are a few references in the other direction which we have already touched on. This leads us to the concept of Supposition. The other lesson is to be found in rules of the scholarly game of Obligation.

It is not necessary or desirable to go at all deeply into the theory of Supposition here: it has been thoroughly discussed by others, particularly by the Kneales (pp. 246–74). A few words, however, will explain its relevance. The theory has quite failed to survive into modern times, principally because it is based on an assumption which would be unhesitatingly rejected by modern philosophers, that the meaning of a term consists solely in its *referring* to certain actually existing things. Modern practice is to distinguish *sense* from *reference*, or *connotation* from *denotation*. Thus the word 'unicorn' is meaningful in spite of the fact that there are no unicorns, and the expressions 'the Morning Star' and 'the Evening Star' have different meanings even though there is only one thing, Venus, to which they both refer.

The supposition of a term, as nearly as we can put it, is what it 'stands for' (*supponit pro*); but a term may stand for different things on different occasions. The word 'man' in 'A man is running' might stand for (the ubiquitous) Socrates, if that is who is being referred to, whereas in 'Man is the noblest of creatures' it stands for the human species or, what is in medieval times the same thing, all men. In 'In every country some man is king' it stands, in a somewhat confused way, for just one man at a time, but a different one depending on which country you have in mind. In '"Man" has three letters' or '"Man" is monosyllabic' it stands (the use of syntactical quote-marks is a modern invention, dating only from Frege[1]) for the *word* 'man', whether as written or as pronounced. Hence in order to sustain the Supposition Theory of Meaning it is necessary to allow that there are several different kinds of Supposition a word may have, depending on context.

We have already seen the beginnings of this theory in Abelard and Adam but it should be obvious that the detailed study of Aristotle's *Sophistical Refutations* provides nearly the full range of distinctions needed: all the examples of Equivocation, and many of the others, can be regarded as cases in which two conflicting Suppositions can be attributed to a word. The names given to the different kinds of Supposition vary somewhat from writer to writer and they are variously classified and divided: the main kinds in William are *material* ('Man is a species'); and *personal*,

[1] *Grundgesetze* vol. 1 (1893), p. 32 of the Furth translation.

subdivided as *determinate* ('A man is running') and *confused*, further subdivided. The word 'man' is confused and *distributive* in 'Every man is an animal'; the concept of Distribution was later to be used in formulating rules for syllogistic reasoning.[1] There are some other subdivisions to deal with difficulties in conjunctive statements and cases of multiple quantification.

A large part of Peter's chapter 6 and of the *parva logicalia*, chapters 8-12, are concerned with Supposition and related concepts. We possess many other such accounts. The theory of Supposition died in the fifteenth century, not to be revived. Nevertheless it was important while it lasted, and some parts of it would stand reanimation.

The game of Obligation is a theoretical development of quite a different order. This, too, has been forgotten; but it has been replaced by nothing else and, although it was never developed at a very high theoretical level, there is every reason to take it seriously and try to learn from it something relevant to modern times.

The complaint that we have made against modern treatments of fallacy, to the effect that, by treating all reasoning as non-dialectical and context-free, they make nonsense of much of the traditional account, applies with almost equal force to William and others of the Middle Ages who eschew dialogue and try to squeeze all reasoning into standard syllogistic form or its minor extensions. We have noticed, however, some signs of strain in William's accounts of Begging the Question and Many Questions, and the closeness to Aristotle of the thirteenth-century tradition guarantees that its exponents cannot be totally blind to the origins in dialogue of much of what Aristotle wrote. It should be added that the exigencies of classroom teaching put them much closer in spirit to Aristotle than we are; for, in a period in which books were handwritten and rare and paper and ink were in short supply, classroom exercises had to be oral. We have seen that many hours of each day of four years of each student's course were somehow spent studying Aristotle's logic, and it is not credible that all this time should have been spent listening to lectures. What did they do? The answer is that they conducted disputations.

[1] See below, ch. 6.

There were probably many kinds of disputation; some, we may surmise, on the Greek model and some more like modern debates with long and measured contributions from speakers in turn. In the case of the *Disputed Questions on Truth* written up by St Thomas Aquinas[1] it is fairly clear that an attempt is made to use the traditional Aristotelian rules of finding arguments to develop syllogisms *pro* and *contra* a given conclusion, perhaps by alternate speakers. The so-called medieval disputations sometimes conducted in modern seminaries are basically of this pattern: compare the Lincoln's Inn debates broadcast in 1949-50.[2] What concerns us is neither of these: it is an elementary classroom exercise whose rules reintroduce Dialectic into Logic and whose workings were, in the long run, to reillustrate much of Aristotle's material.

One of the earliest – perhaps the earliest – of the treatises on Obligation is ascribed to William of Sherwood. The ascription cannot be entirely certain. However, there are no extant treatises earlier than the time of William. Several dozen treatises from the fourteenth century or later are known.

As with the Greek debates, the Obligation game was played by two people, here known as the *opponent* and the *respondent*: the first of these would usually have been the teacher.[3] Also as in the Greek debates, the respondent is committed to uphold a thesis, known as the *positum*; and submits to the equivalent of direct questions in so far as propositions, called *proposita*, are put to him by the opponent for his assent or dissent. With this the

---

[1] Thomas's *Disputed Questions*, which reads as if it might have been inspired by (though it is by no means a literal transcript of) actual classroom discussions, was in an established literary tradition. The earliest example is the *Questions on Holy Writ* of Robert de Melun (*c.* 1145); and in the same tradition is the *Disputations* of Simon de Tournai (*c.* 1201). See Chenu, *Introduction à l'étude de s. Thomas d'Aquin*, pp. 67–77.

[2] One of these, Cusack and others, 'The Cinema is the Highest Form of Art', broadcast 23 January 1950, has been published. There had been an earlier one on a more scholastic subject the previous October.

[3] In dealing with Obligation, which has not been much discussed in modern times, I have relied heavily on an unpublished thesis by Romuald Green, O.F.M., *An Introduction to the Logical Treatise "De Obligationibus"; with critical texts of William of Sherwood (?) and Walter Burley*, presented at the Catholic University of Louvain, 1963. I am also indebted to personal discussions with Fr Green.

resemblance ends; and, although the features already mentioned are enough to make it likely that the origin of the game is to be found in Aristotle's *Topics* and *Sophistical Refutations*,[1] its development has given it a form of its own.

The *positum*, in the first place, is not necessarily a thesis to which the respondent personally subscribes: it is handed out by the opponent, and is generally a *false* statement, and understood as such by both opponent and respondent. It is usually, also, quite simple in content – that Socrates is black, that the respondent is a bishop, or that Man is not an animal – unlike the abstract theses of the Greeks, such as that virtue is teachable. Having accepted the false *positum* the respondent must be quite prepared to be led into other falsehoods to maintain consistency, suspending his own beliefs; but he must tell the truth in all cases in which he can do so consistently with the *positum* and such other propositions as he has already admitted. Suppose, for example, the *positum* is

Socrates is black

and the two *proposita* are

[1] Plato is black, and

[2] Plato and Socrates are the same colour.

The first *propositum* must be denied, simply because it is false and nothing else hangs on it; but the second, though really true, must now also be denied, for the sake of consistency. If the *proposita* had been presented in the opposite order they would both have had to be conceded, by similar reasoning.

The game finishes by the opponent's saying *Cedat tempus*, usually when the respondent has been trapped in a contradiction or when he has clearly succeeded in avoiding such a trap.

In variants of the procedure, the respondent may be obliged to deny, rather than affirm, a given thesis, or to hold a given thesis as doubtful. It is not clear that the task of denying a thesis is different from that of upholding its contradictory, though it is in theory possible, of course, that it should be so interpreted. An elaboration of the game is that the opponent, besides laying

---

[1] Green (p. 26) notices that part of the opening paragraph of Burley's treatise is nearly verbatim from the opening of Book VIII of the *Topics*; and comments on a number of lesser connections.

down a *positum*, may specify certain statements as 'actual fact' (*rei veritas*). This is different from including them with the *positum* since the respondent is not obliged to maintain them if they are inconsistent with the *positum* or whatever else he admits. The result is a two-level pretence on the part of the respondent. Thus we could adapt the previous example by writing 'white' for 'black' and vice versa, subject to a specification that, in 'actual fact', Socrates and Plato are both black.

'Sophisms' occur when unexpected or paradoxical results are obtained. Rules are given for constructing cases of interest. In the self-explanatory words of the treatise attributed to William (translated from Green, vol. 2, p. 11):

> One way of constructing sophisms is: to assume a conditional whose consequent is false and construct a disjunction of the consequent and another proposition, and another disjunction of the antecedent and the opposite of the proposition; then to conjoin those disjunctions [as *positum*]; and then, *proposit* that the consequent is false. For example: In actual fact, Socrates does not exist. Let A mean 'Socrates is running or you are standing', and let B mean 'Socrates is moving or you are not standing'. Conjoin A and B [as *positum*]. Then [*propositum*]: 'Socrates is moving'. This should apparently be denied, since it is false and does not seem to follow. If it is denied, let 'Socrates is running' be *proposited*. This must be conceded, because it follows: thus, if A and B are true and Socrates is not moving, B is true on account of the truth of 'You are not standing'. Therefore, A is true *not* on account of 'You are standing' but on account of 'Socrates is running', and so this must be conceded. This done, 'Socrates is moving' is again *proposited*. If you concede it, *cedat tempus*, you have conceded and denied it in the same disputation, which is wrong. If you deny it, *cedat tempus*, you have denied something that follows, which is wrong.

The solution, though it is not given in the text, is that 'Socrates is moving' really does follow from the *positum*, and should never have been denied in the first place. Another example is worth giving, to illustrate the complications of the theory in the case in which the respondent undertakes the obligation to hold a proposition (the *dubitatum*) as doutful (p. 33):

> But the following is a sophism concerning doubting: In actual fact, Socrates does not exist. It is to be *doubted* that Socrates is running.

Now [first *propositum*]: Socrates is moving. If you deny this, let it be *proposited* that Socrates is running. It is now necessary to deny this too; but it is the *dubitatum*. If you concede it, *cedat tempus*, you have conceded a falsehood without being obliged to do so, which is wrong. If you reply that it is doubtful the objection is the same, because to doubt an antecedent does not oblige you to doubt a consequent. (If something is seen walking in the distance it is possible to doubt whether it is a horse, yet know that it is an animal.)

It should be added that this is the case if the consequent is really true, but if the antecedent is really false and is doubted the consequent is to be doubted. Thus it is not known to be true, since it is false, and it is not known to be false, because then it would be known that the antecedent was false. In this case, then, a doubtful reply is appropriate to 'Socrates is moving'.

William's(?) analysis is not very good at this point since the solution he favours is one against which he has just given a convincing argument. However, even a modern logician might need to resort to symbols to do better.[1]

Very many of the examples in treatises on Obligation turn on self-reference. Thus let the *positum* be 'You have not replied to the *propositum* "God exists"'. If 'God exists' is now *proposited* it must be conceded; but this has made the *positum* vulnerable, and William(?) says the *positum* has become 'impossible *per accidens*'. We have set cases of this kind outside our terms of reference and need not linger over them. They illustrate, however, the extent of the preoccupation of medieval writers and teachers with the problems raised by Dialectic in the older, Platonic, sense. Once again, very little progress has been made in integrating the new developments with the inherited Logic. It can now be shown, however, that a start was made here and there towards reapplying the Obligation game to the material of the traditional accounts of fallacies.

Teachers conducting classes in Obligation needed lists of

[1] The appropriate logic is isomorphic to a deontic one as in, for example, von Wright, *An Essay in Modal Logic*, ch. 5. Let p be 'Socrates is running' and let q be 'Socrates is moving': interpret 'O' as 'The respondent is committed to saying that', and assume $O(p \supset q)$. The obligation in respect of the *dubitatum* is represented by $-Op$. $-O-p$. When q is *proposited* the respondent must resolve the situation with Oq, $O-q$ or $-Oq.-O-q$. There are convincing objections to $O-q$, but he can reply with a disjunction of the others, equivalent to $-O-q$, a possibility William(?) has not envisaged.

examples to work from, and we possess many such lists, dating from as early as the twelfth century (well before William) to the late fourteenth or even the fifteenth century.[1] Treatises on 'sophisms' also became a literary form in which solutions and objections were discussed. Perhaps the most interesting of these treatises is that of Buridan[2] (about 1330). Another is that of his pupil Albert of Saxony (1316?–90?)

The content of the examples in these lists reflects the predominant interests of medieval logicians and, if any generalization is possible, it would be that they are concerned with Supposition theory. However, this is only part of the truth. Some of them, particularly in the earlier lists, are simple exercises in logical consequences; many involve paradoxes of self-reference; some deal with sophisticated philosophical issues; and, in the later period, some lists deal almost exclusively with physical science. There is a traditional classification of the examples into *insolubilia*, *impossibilia* and *sophismata* but it is not always a very fruitful one in detail.

Even the twelfth-century lists use some Obligation terminology, showing that the Obligation game was practised half a century or more before the earliest treatises: examples often begin with some such phrase as *Facta tali positione* ('It is *posited* that'), and the discussions use *cedat tempus* to indicate the end of a disputation through the breaking of a rule. Some specimen sophisms from an early list are (Grabmann, 'Die Sophismataliteratur', pp. 21–3)

> Every man and every ass are two.
>
> Nothing differs from nothing.
>
> Opposites and non-opposites are opposites.
>
> It is *posited* that Socrates killed Plato's father and Plato killed Socrates's father, and it is asked whether the following is true: Socrates and Plato killed their fathers.
>
> It is *posited* that every man is running and no ass is running, and it is asked whether the following is true: Every man or ass is running.

---

[1] See particularly Grabmann, 'Die Sophismataliteratur des 12. und 13. Jahrhunderts'.
[2] *Sophismata*: Buridan also gave an Aristotelian treatment of Fallacies in his *Compendium Totius Logicae*, tract 7.

It is *posited* that a certain layman had that man as a son and was afterwards made a priest, and it is asked whether the following is true: That man is the son of a priest.

The thing about which I am not speaking is something.

There are about fifty sophisms in the list from which these are taken.

Later in the thirteenth century many of the sophisms reflect the growing nominalist movement, and Siger of Brabant in 1277 was even called to Rome to explain his support of the sophism 'All men are animals, even if no men exist', which depended on the heterodox premiss that the concept Man would still exist even after the Day of Judgment. Peter of Spain was by now pope, and it would be interesting to know what kind of discussion they had. (Siger was put in the charge of a 'secretary', who later went mad and killed him.)

A very few Aristotelian fallacies creep into the early lists: 'The white thing can be black' is fairly common. In a manuscript of the late thirteenth century which seems to contain the minutes of some kind of international Conference of Sophismatists (Grabmann, 'Die Sophismataliteratur', pp. 70-2): we find 'Whether there are five fallacies *in dictione*, no more, no less' as a sophisms allegedly resolved by Master Petrus de Insula. At the same meeting Johannes de Alliaco contributed the solution of 'Whether the proof of *Something exists* is intermediate' (that is, neither intrinsic nor extrinsic) and Petrus de Bognovilla resolved 'Whether there can be a science of proof'. That these discussions were regarded as being in the *sophismata*-tradition is an indication of the development and deepening of the subject.

That the *Sophismata* of Buridan draws heavily on the background of the Obligation game can be seen not only from the use of language but also from some explicit discussions; for example, the discussion of sophism (3) of chapter 8 (ed. Scott, p. 185), 'If every man runs, then an ass runs' where what is debated is the admissibility of *positing* a 'necessary' falsehood such as 'Every man is an ass'. Buridan is strongly nominalist, in the tradition of Ockham, and continually produces examples such as 'Every spoken proposition is true' whose 'proof' rests on – and hence demolishes – the theory that every spoken proposition signifies a mental proposition and 'howsoever it signifies, so it is correspondingly in the thing signified' (p. 64). The realization comes as a surprise that one of these examples, 'No man lies' (p. 70), is straight out of Plato's *Euthydemus*; though this is probably not a case of direct borrowing at all, but a re-invention. Buridan adapts to his medium several of the Aristotelian examples from treatises on fallacies, fitting them into his own classification scheme. Chapter 4 (pp. 110–24) which is 'about Connotations', has (I give only the opening arguments):

> [2] *You ate raw meat today*. I posit the case that you bought a piece of raw meat yesterday, and today you ate it well-cooked...
>
> [3] *The white will be black*...
>
> [6] *I saw Peter and Robert*. Posit that I saw Peter yesterday and Robert the day before yesterday.... The opposite is argued, since ... it was never true to say, in the present tense, 'I see Peter and Robert'...
>
> [7] *This dog is your father*. This is proved, because this is a father and this is yours; hence, it is your father. Similarly, pointing at a black monk it is argued that there is a white monk, because this is white and it is a monk;...
>
> [9] *You know the one approaching*. I posit the case that you see your father coming from a distance, in such a way that you do not discern whether it is your father or another...

In the case of the 'raw meat' example Buridan is launched into a discussion of substances and properties. Yesterday, I bought 'not only the substance of the meat, but also its accidents and properties'. Certainly, 'in cooking, some of the substance of the meat disappears and evaporates', but this is not the main point since, even if the substance were unaltered, the properties could still be altered. There are lengthy discussions of the other examples. To [9] the general remark is made that the verbs 'understand', 'be acquainted with', 'know' and others (p. 126):

> effect some special kinds of connotation in the terms with which they occur. For since I can understand the same thing in many different ways, and, corresponding to these diverse ways, impose different names on it to signify it, therefore, such verbs cause the terms with which they occur to connote the reasons for which their names are imposed to signify them, and not only the external things known, as happens with other verbs.

THE ARISTOTELIAN TRADITION 133

There are isolated other treatments of Aristotelian examples. The ultimate nominalist thesis is examined in the sophism 'It is in our power that a man is an ass' (p. 164):

> It is proved, since you and I shall argue, and then we can use conventional words in whatever way we agree to...

A few of Buridan's sophisms raise issues in physical science – 'All which is moved was moved previously', 'No change is instantaneous' – and, soon after Buridan's time, 'physical' sophisms were very common. Particularly to be mentioned are the Merton College (Oxford) group of the 1330s and 1340s, William Heytesbury, Thomas Bradwardine, Richard Swineshead and, a little later, John Dumbleton and Richard Kilmington.[1] Exegesis of the material of Aristotle's *Physics* and *The Heavens* had been common earlier, in William of Sherwood and others; but the Merton writers produce elaborate puzzles concerning what we can only describe as the mathematics of beginning and ceasing, maxima and minima, change, motion and velocity. To take just one example: Socrates is one foot long and Plato is two feet long; both increase uniformly in length for one hour, at such a rate that at the end of the hour both would be exactly three feet long; but, at the final instant of the hour, each of them ceases to exist. Heytesbury (Wilson, p. 47) makes it clear that he means by the latter *positum* that the final instant of the hour is to be the first instant of their non-existence rather than the last instant of their existence, so that Plato and Socrates never 'begin to be' the same size; but he thinks that it is nevertheless correct to say 'Socrates and Plato will be equal'. In the process, the mathematical theory of bounds and limits gets a good airing.

The function of the *positum* in these examples has changed subtly and we have now not so much an Obligation game as a thought-experiment. It should be added that Galileo, though he had no high regard for medieval logic, refers to Heytesbury and was influenced by him. Perhaps the greatest thought-experiment in the history of science was that of Galileo. The form is dialogue:[2]

---

[1] See Wilson, *William Heytesbury: Medieval Logic and the Rise of Mathematical Physics*. For manuscript sources see the bibliography in this work.
[2] Galilei, *Dialogues Concerning Two New Sciences*, p. 63.

SALVIATI If then we take two bodies whose natural speeds are different, it is clear that on uniting the two, the more rapid one will be partly retarded by the slower, and the slower will be somewhat hastened by the swifter. Do you not agree with me in this opinion?

SIMPLICIO You are unquestionably right.

SALVIATI But if this is true, and if a large stone moves with a speed of, say, eight while a smaller moves with a speed of four, then when they are united, the system will move with a speed less than eight; but the two stones when tied together make a stone larger than that which before moved with a speed of eight. Hence the heavier body moves with less speed than the lighter; an effect which is contrary to your supposition.

The hypothesis, that the 'natural speeds' of bodies are independent of their size, did not need to be tested empirically from the Tower of Pisa. The style in sharp contrast with what has preceded it, is Plato's; but the argumentation is unmistakably medieval.

CHAPTER 4

Arguments 'Ad'

What happened to the Middle Ages? Surveying the two centuries preceding 1500 we could be pardoned for thinking that in 1350 logicians were all suddenly carried off by plague. A few slightly-known names can be found after this – Paul of Venice, Paul of Pergula, Ralph Strode, George of Trebizond – but the overall picture is nearly blank. It has been customary to trace some part of the change of direction of academic thought to the invention of printing, but though this may have affected the later revival it can hardly explain the dearth of original thought in the century or so before 1450. Again, the scholars who left Constantinople when it fell to the Turks may have stimulated European learning when they arrived, but we can hardly explain the ossification of Logic before 1453 by the fact that Constantinople was still in Byzantine hands. The non-discovery of the New World cannot be blamed for things that did not happen before 1492. Did the Hundred Years War really so upset the scholars of England and France? If the Middle Ages are defined as the period of the pre-eminence of Logic, they stopped well before the Renaissance took over.

Whatever the explanation, although the medieval logical synthesis continued to be at the base of university studies, it clearly became progressively rigid and fleshless in the absence of new stimulus. Sir Thomas More, writing in 1516 about the happy inhabitants of Utopia, said (ed. Surtz and Hexter, p. 159):

> ... in music, dialectic, arithmetic, and geometry they have made almost the same discoveries as those predecessors of ours in the classical world. But while they measure up to the ancients in almost

all other subjects, still they are far from being a match for the inventions of our modern logicians. In fact, they have discovered not even a single one of those very ingeniously devised rules about restrictions, amplifications, and suppositions which our own children everywhere learn in the *Small Logicals*. In addition, so far are they from ability to speculate on second intentions that not one of them could see even man himself as a so-called universal . . .

The 'Small Logicals' are the *parva logicalia* in the second half of Peter of Spain's *Summulae*, and could be stretched to include Obligations, Insolubles, and the non-Aristotelian Sophisms. Sir Thomas could hardly have made this judgement if he had lived two centuries earlier, but it was fair comment on what survived of the tradition among his contemporaries. The Obligation doctrine was particularly badly served by inferior printed treatises.[1]

The subsequent history of the study of fallacies – from the Renaissance to the present – is a series of waves of anti-Aristotelian attempts to get rid of the subject altogether, followed at regular intervals by the reinstatement of the old doctrine in ever new revised forms. In the process we have drifted far away from Aristotle; but as fast as one logician has declared him redundant, another has come forward to re-employ him, at least in theory. There have been three groups either actively opposed to Fallacies or uninterested in them: the first, Agricola and Ramus; the second, Locke and the empiricists, matched by Leibniz and the rationalists; the third, Boole, Frege and Russell. Answering revivals have been instituted by, first, Fraunce, Buscher, Bacon and Arnauld; secondly, Whately, J. S. Mill and De Morgan; thirdly, the modern, mainly American, logicians that have been cited extensively in chapter 1. A century off, a century on: the cycle is not, of course, quite so regular as this, but it is more pronounced than most historical patterns.

There was not, at first, any open anti-Aristotelianism. We know the fourteenth and fifteenth centuries as the age of Dante, Petrarch, Boccaccio, Chaucer and Villon, when literature stole the scene and made the decaying scholastic tradition seem lifeless. Towards the end of the period we begin to find a renewed inter-

---

[1] See the treatises on Obligation and Insolubles appended to Peter's work in Peter of Spain, *Tractatus Syncategorematum and Selected Anonymous Treatises*. These are from editions of 1489 and 1494.

est in Rhetoric, conceived as a science of literary style. The most important figure of this movement is Rodolphus Agricola (or Roelof Huusman, 1444–85) who, besides writing his *Dialectical Invention*, was a painter and musician – 'a sort of minor Leonardo da Vinci'[1] – and who tried to unify Dialectic and Rhetoric by taking the emphasis away from deductive reasoning and putting it on Topics. This reversion to early Aristotelianism was not accompanied by an interest in fallacies, which, significantly, are not mentioned at all. The combined subject is divided into Invention and Judgement, the latter being identified with Disposition or Arrangement, following Cicero's conception of Rhetoric. This set a fashion which was followed, with variations, by most logicians for the next two hundred years and even influenced theories of experimental method in the emerging sciences.

Agricola's opposition to the traditional Aristotelianism was implicit rather than open, but he initiated a movement that was to take some time to develop. His contemporaries were generally more traditional than he. A treatise on fallacies under the rather surprising authorship of Girolamo Savonarola (1452–98), the monk who assumed leadership of the reform movement in Florence, is less revolutionary than its writer's politics. Equally typical of its time, notable in this case for literary rather than logical originality is the *Margarita Philosophica* ('Pearl of Philosophy', 1496) of Gregor Reisch (d. 1525), which was an eight-volume philosophical encyclopaedia *in verse*, and assumed enormous popularity, running through nearly as many editions as Peter of Spain's *Summulae*. A little later, another noted reformer whose energies went elsewhere than into the reform of Logic was Philipp Melanchthon (1497–1560; the name is a graecization of Schwarzerd) colleague of Martin Luther: his *Commonplaces* is one of the classics of Protestant theology, but his *Dialectical Questions*, a logic text in the form of a catechism, is unremembered. It shows Agricola's influence, but reintroduces a traditional treatment of fallacies.

For a serious attempt to overthrow the existing order we must move ahead to Petrus Ramus (1515–72), who became famous in

[1] So described by Ong in his *Ramus: Method and the Decay of Dialogue*, pp. 95–6. Agricola's work was completed about 1479 and ran through many editions in the sixteenth century.

1537 for his dissertation-thesis 'Everything Aristotle said was false'. The impiety of this pronouncement would hardly have shocked academics familiar with medieval sophisms, but it seems to have impressed the public and he dined out on it all over Europe in a short career that brought him widespread, if undeserved, popularity. There is hardly any philosophical method more unpromising than that of allowing one's opponents to ask the questions and disagreeing with their answers. Ramus's positive logic is a deductive theory based, following Agricola and Melanchthon, on a set of topics of reasoning.[1] The study of fallacies is dismissed as unnecessary:

> First, should not the overall description of vices itself arise from the direct opposition of virtues so that, for every kind of virtue there should be just one kind of vice? And in so far as there are two overall virtues in dialectic, one of Invention and one of Judgment, so there ought to be two overall vices, one opposed and hostile to true Invention, the other to correct Judgment; so that, to the virtues of sensible discourse there are opposed the contrary vices of deception, and captious argument to true, the faulty arrangement of what is invented to the correct and constant. But how could you expect light from the author of darkness?

This is his regular title for Aristotle. A little further on he continues:

> What is it that is sold in the shops of the Aristotelians? Fictitious and fabricated wisdom? Surely not. Then what? Madness covered with the false simulation of wisdom?...
>
> How many kinds of sophisms are there? There are five: refutation, falsification, paradox, solecism and babbling. There are six divisions of refutation *in dictione*, of which five – equivocation (in the terminology of the Aristotelians), amphiboly, composition, division and accent – are sufficiently comprehended in one word, in the fault of *ambiguity:* which is a common fault of all speech, and does not need these empty absurdities of subdivisions. Figure of speech is captious similarity; but an incomplete similarity, for it can also be a fallacious one. There are seven kinds of fallacies *extra dictionem*, the fallacy of accident, *secundum quid, ignoratio elenchi*, begging the question, consequent, non-cause as cause, diverse

---

[1] Ramus's earlier, and most anti-Aristotelian, works are the *Dialectica institutiones* and *Aristotelicae Animadversiones* (both 1543). The quoted passages are from the latter, ff. 70–2.

question. When Aristotle reckons these seven kinds to be contained within *ignoratio elenchi*, we shall believe the man: the rest we shall laugh at. Yet I see begging the question stuck in in third place: this excresence enormously pleases the Aristotelians, and it seems enormously appropriate that it will be made sport of by such refined intellects...

The fallacy of *secundum quid* is a false species of division, for the whole seems to be equated to a part: the source of the fallacy of non-cause as cause is indicated by its name: the fallacy of the consequent is a captious kind of related secondary syllogism: the fallacy of accident, of diverse questions, of *ignoratio elenchi* do not stand in need of new rules...

Of the remainder, the ostentation of falsification, paradox, solecism and babbling is not so much obscure as ridiculous.

Elsewhere he says that the *Organon* is not authentic but was attributed to Aristotle by some sophist enemy of truth and science.[1] The superficiality of these unargued pronouncements is matched only by their historical importance. This kind of criticism of current Logic was overdue, and led many men more worthy than their author to call themselves Ramists. He became a martyr by being killed in the riots of St Bartholomew's Eve in 1572, and was celebrated in Christopher Marlowe's *Massacre at Paris*.

From this time, editions of medieval and traditional authors stop, and fresh new books begin to appear. In 1551 Thomas Wilson published his *Rule of Reason*, the first of these and, incidentally, the first logic book in English. (Ramus's *Dialectique* of 1555 was the first in French.) He was followed by a succession of some dozen 'Elizabethans',[2] nearly all writing in English, and in varying degrees of allegiance to Ramus. It is common for the subject to be laid out as a series of Topics, and for attitudes of Ramus to be echoed. In one respect, however, these writers are united: they are unwilling to forego the opportunity of writing on fallacies.

One of the most interesting of the Elizabethans is Abraham Fraunce, whose *Lawiers Logike*, though in English, is larded with

[1] See discussion by Waddington, *Pierre de la Ramée*, p. 366.
[2] Howell, in his *Logic and Rhetoric in England, 1500–1700*, has notes on logic books by Thomas Wilson, Ralph Lever, Abraham Fraunce, Thomas Granger, Thomas Blundeville, Samuel Smith, Robert Sanderson, Christopher Airay, John Newton and some others.

excursions into Latin and what he calls the 'hotchpot French' of England's old legal statutes. He uses legal case-histories extensively as examples. He is a little apologetic about his inclusion of notes on fallacies (f. 26):

> Sophistry, as I have said elsewhere, is no Logike: therefore least I should injury the art by joyning sophisticall fallacians with Logicall institutions, I have rather reserved them to these annotations, then thrusted them in among the precepts. Some use, I confesse, there may bee had of them. . . . But if wee shall put downe every thing in Logike, which hath any little shew of profite thereunto; Grammer will be good Logike, because it helpeth us to utter yt which we have Logically conceaved.

This said, the 'annotations' can proceed. New kinds of fallacy are discussed, though many traditional ones are included. The arrangement is also a mixture of old and new, as may be seen from the attached figure. Those shown are all presented and discussed early in the book, when the concept of a Fallacy is first introduced, with the exception of those 'belonging to Disposition', which are discussed only briefly in a later section, and those 'peculiar to certain places': these are numerous, and discussions of them are appended to discussions of the various Topics or 'places' to which they are supposed to be related. Thus under the Topic of 'Cause', representing argument from cause to effect, we have mention of the Fallacy of 'arguing from that which is no cause: as if it were a cause'; and under the Topics of 'Whole, part, generall, speciall' we have the Fallacies of Composition, Division, and Whole and Part. Some of Fraunce's examples have been given in chapter 1.

So the Ramist revolution both succeeded and failed: Logic took on a new shape but remained nearly unchanged in content. Perhaps the most significant change – though it does not, at first seem so as we read the writers who made it – was the introduction into Logic, from the rhetorical tradition, of the 'extrinsic' or 'inartificial' arguments of Aristotle and Cicero. These appear in Ramus,[1] and were taken up by most of the Elizabethans. Abraham Fraunce calls them 'Secondary arguments' and has a rather strange extended list of them, including Distribution or

---

[1] *Dialectique* (1555), pp. 96–101.

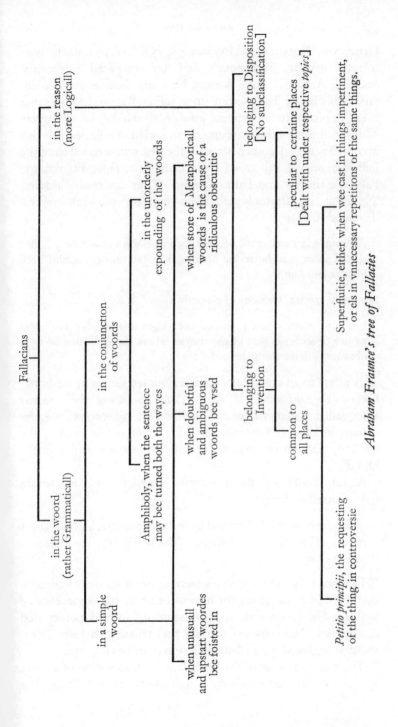

Abraham Fraunce's tree of Fallacies

Definition, Notation or Etymology, 'Of the argumente borrowed' (namely, Testimony), and 'Of compared arguments' (namely, Analogy). Like Ramus, he treats them as supplementary to his list of Topics; but since, unlike Ramus, he holds that every Topic has its own special kind of 'elench' he also effectively adds the abuses of these arguments to his list of fallacies. For this involved reason, Fraunce becomes the first writer to tell us of the Fallacies, among others, of False Definition, False Etymology, False Testimony and False Analogy. Thus under the heading 'Notation or Etymologie' there appears the sub-heading 'Elenchs' (f. 56):

> A Woman is a woe man, because shee worketh a man woe. . . . But all the sport is to heare the Moonkish notations of woordes both Greeke and Latine. . .

Similarly, under 'Elenchs of all definitions' (f. 64):

> First, if it want where it should bee: Then if it bee, but bee false: or bee obscure, as that of the snayle, where the definition is more obscure than the thing defined. . .

This refers to an example of Cicero's quoted earlier in the book. Cicero (*De Divinatione*, II, 64, 133) makes fun of the scientist who called a snail *terrigenam, herbigradam, domiportam, sanguine cassam*; in Fraunce's translation 'such a beast as is bred of the ground, walketh on grasse, carrieth her cottage, and wanteth bloud'.

Again, under 'Of the argument borrowed' we once more have 'Elenchs' (f. 71):

> A false testimonie is descried by the wickednesse, and malicious nature of him that gave witnesse. The world is full of false forsworne knaves. . . .

If Fraunce's thought-processes were those of modern logicians, this would be the signal for him to bring in the *argumentum ad hominem*. He does not do so, but has nearly, instead, perpetrated one himself. Nevertheless the fact that reliance on False Testimony is regarded as an Elench is a sign of things to come.

The ultimate irony of the times was the publication of a book specifically devoted to giving a Ramist theory of fallacies. We have mentioned Heizo Buscher's treatise above.¹ It does not possess any merits that would warrant our discussion.

The greatest of all Ramists, however, was Francis Bacon (1561–1626), prosecutor of Essex on behalf of Elizabeth I, Lord Keeper of the Seal and Lord Chancellor of England under James I, famous for his visionary championing of the cause of organized science. He was, perhaps, less a logician than any other person whose writings are discussed in this book; he was a lawyer and politician. The physician William Harvey, who knew him and sometimes attended him in a professional capacity, said of him 'He writes Philosophy like a Lord Chancellor'. De Morgan remarks (*A Budget of Paradoxes*, vol. 1, p. 79):

> This has been generally supposed to be only a sneer at the *sutor ultra crepidam* [cobbler not sticking to his last]; but we cannot help suspecting that there was more intended by it. To us, Bacon is eminently the philosopher of *error prevented*, not of *progress facilitated*. When we throw off the idea of being *led right*, and betake ourselves to that of being *kept from going wrong*, we read his writings with a sense of their usefulness, his genius, and their probable effect upon purely experimental science, which we can be conscious of upon no other supposition. It amuses us to have to add that the part of Aristotle's logic of which he saw the value was the book on *refutation of fallacies*. Now is this not the notion of things to which the bias of a practised lawyer might lead him? ...

This is unjust in that it suggests small-mindedness. In fact, Bacon was forever drawing up large schemes, political and literary, restlessly recasting his ideas in new forms. The early *Advancement of Learning* (1605) was ambitious in its attempt to survey the entire state of learning in his day: the later *New Organon* (1623), whose title is enough to indicate his logical aspirations, was still only a part of the plan to end all plans, the *Great Instauration* (or, perhaps, *Restoration*), which was planned to include virtually all his repetitious writings and much more besides. It hardly needs saying that it was never finished.

He has left us half a dozen overlapping treatments of his famous Idols, inconsistent in various ways and fitting irregularly into different overall patterns. They are first described, unnamed, in the *Advancement*. Surveying Logic, he finds that the

¹ p. 10.

study of sophisms is, on the whole, in a satisfactory condition (*Works*, vol. 3, pp. 393 ff.):

> This part, concerning Elenches is excellently handled by Aristotle in precept, but more excellently by Plato in example, not only in the persons of the Sophists, but even in Socrates himself; who, professing to affirm nothing, but to infirm that which was affirmed by another, hath exactly expressed all the forms of objection, fallace and redargution...

However, there is more to Elenches than this.

> ... But lastly, there is a much more important and profound kind of fallacies in the mind of man, which I find not observed or enquired at all, and think good to place here, as that which of all others appertaineth most to rectify judgment: the force whereof is such, as it doth not dazzle or snare the understanding in some particulars, but doth more generally and inwardly infect and corrupt the state thereof.

This is a general statement, but we now get down to detail.

> For this purpose, let us consider the false appearances that are imposed upon us by the general nature of the mind, beholding them in an example or two; as first, in that instance which is the root of all superstition, namely, *That to the nature of the mind of all men it is consonant for the affirmative or active to affect more than the negative or privative:* ...

The passage which follows has already been quoted above.[1] These 'false appearances' are what Bacon later referred to as the 'Idols of the Nation or Tribe'. The word 'idol' ($\epsilon\ddot{\iota}\delta\omega\lambda o\nu$) means 'false appearance' and, in the seventeenth century, had overtones of irreligion. In a discussion of religion near the end of the *Advancement* Bacon defines 'idolatry' as 'when we worship false gods, supposing them to be true', echoing the Aristotelian definition of fallacy and revealing that he already associates the one field with the other. Idols were things that were falsely worshipped. The name 'Nation' or 'Tribe' is not very clear in this connection but is explained in the *New Organon* (vol. 4, p. 54: in this case the text is a translation from Bacon's Latin):

> The Idols of the Tribe have their foundation in human nature itself, and in the tribe or race of men. For it is a false assertion that the

[1] p. 47.

sense of man is the measure of things . . . the human understanding is like a false mirror, which, receiving rays irregularly, distorts and discolours the nature of things by mingling its own nature with it.

The *Advancement* next moves to other cases (vol. 3, p. 396):

> Let us consider again the false appearances imposed upon us by every man's own individual nature and custom, in that feigned supposition that Plato maketh of the cave: for certainly if a child were continued in a grot or cave under the earth until maturity of age, and came suddenly abroad, he would have strange and absurd imaginations; so in like manner, although our persons live in the view of heaven, yet our spirits are included in the caves of our own complexions and customs; which minister unto us infinite errors and vain opinions, if they be not recalled to examination.

These are 'Idols of the Cave', and the reference is to the allegory in Book IX of Plato's *Republic*. Next:

> . . . lastly, let us consider the false appearances that are imposed upon us by words, which are framed and applied according to the conceit and capacities of the vulgar sort: and although we think we govern our words, and prescribe it well, *Loquendum ut vulgus, sentiendum ut sapienties*, [a man should speak like the vulgar, and think like the wise;] yet certain it is that words, as a Tartar's bow, do shoot back upon the understanding of the wisest, and mightily entangle and pervert the judgment; . . .

In his early work *Valerius Terminus* Bacon called these 'Idols of the Palace' (vol. 3, pp. 242, 245), but the name might have been insulting to James and he changed it to 'Idols of the Market-Place'.

The fourth kind, 'Idols of the Theatre', did not occur in the *Advancement* and in order to find their origin in his early work we must look at a short, illuminating if undergraduate piece with the cryptic title *Temporis Partus Masculus*, 'The Male Birth of Time' (vol. 3, pp. 528–39): originally he had called it *Temporis Partus Maximus*, 'The Greatest Birth of Time'. This is a speech from the bench at the trial for incompetence of a number of history's leading philosophers, and is slightly reminiscent of Lucian. The judge heaps invective on the accused in a way which is quite uncharacteristic of Bacon in his other writings and, from what we know, in real life: he was normally both generous and balanced. Here, his deep anti-intellectualism is revealed, when Aristotle is described as 'that worst of sophists, stupid with useless subtlety, a cheap mockery of verbiage'. He never quite lost this attitude. Now, in the *New Organon*, he writes (vol. 4, p. 55):

> Lastly, there are Idols which have immigrated into men's minds from the various dogmas of philosophies, and also from wrong laws of demonstration. These I call Idols of the Theatre; because in my judgment all the received systems are but so many stage-plays, representing worlds of their own creation after an unreal and scenic fashion. Nor is it only of the systems now in vogue, or only of the ancient sects and philosophies, that I speak; for many more plays of the same kind may yet be composed and in like artificial manner set forth; ...

The detailed discussion that follows gives more measured criticisms of some particular philosophers, using material from his earlier *Refutation of Philosophies* (1608-9).

The Idols have seldom found their way explicitly into books on Logic but they summarize more completely than any other piece of work a changed attitude towards fallacy, sophism, and error. From now on, some part of the analysis of fallacy will involve an appeal to psychological factors, as in the Idols of the Tribe and Cave, or to social ones, as in the Idols of the Market-Place. The Idols of the Theatre are invoked less directly but, at least in British empiricist tradition, an argument from authority is now nearly always considered to be a fallacy, rather than extrinsically valid; and many are in sympathy with Hume's advice as to the disposal of any book found to contain neither empirical fact nor reasoning concerning quantity or number: 'Commit it then to the flames, for it can contain nothing but sophistry and illusion'.[1]

In passing we might notice that Sextus Empiricus had something to do with forming these attitudes. Sextus had not been entirely unknown to the Middle Ages, and a Latin translation of *Outlines of Pyrrhonism* is found in one thirteenth-century manuscript.[2] In the sixteenth century, however, Sextus and Pyrrhonism were 'discovered' in circumstances in which they could make the greatest impact, and, for a time, 'the divine Sextus', as Le

---

[1] Hume, *Inquiry* (1748); final sentence.
[2] Paris, Bibliothèque Nationale, MS. Fonds Latin 14700, ff. 83-132.

Vayer called him, even seemed to overshadow Aristotle.[1] Montaigne's long essay 'An Apology of Raymond Sebond' (1580) is one of the classic texts of this revival and an English translation of it appeared in Bacon's time.[2] But though there is no doubting Sextus's influence in spreading philosophical scepticism, his critique of the concept of fallacy made no discoverable impression: it is not mentioned at all by Montaigne.

Before leaving Bacon we should note that he also had something to say about sophistical stratagems; in his essay 'Of Cunning' (vol. 6, pp. 428-31):

> Another [point of cunning] is, that when you have anything to obtain of present despatch, you entertain and muse the party with whom you deal with some other discourse; that he be not too much awake to make objections. I knew a counsellor and secretary, that never came to Queen Elizabeth of England with bills to sign, but he would always first put her into some discourse of estate, that she mought the less mind the bills.

and again

> There is a cunning, which we in England call *The turning of the cat in the pan*; which is, when that which a man says to another, he lays it as if another had said it to him. And to say truth, it is not easy, when such a matter passed between two, to make it appear from which of them it first moved and began.

This is the model for future conceptions of sophistry: it is 'a sinister and crooked wisdom'. Bacon's essay makes no pretence of systematic or theoretical treatment:

> But these small wares and petty points of cunning are infinite; and it were a good deed to make a list of them; for that nothing doth more hurt in a state than that cunning men pass for wise.

The *Essays* predate the *Advancement*; and this programme for a new *Sophistical Refutations* – if it was ever really Bacon's own – was rewritten when the *Advancement* was undertaken.

Bacon's most outstanding philosophical disciple was Hobbes (1588–1679) who, however, though even less a logician than his

---

[1] See Popkin, *History of Scepticism from Erasmus to Descartes*, p. 17.
[2] Montaigne, *Essays*. John Florio's translation of 1603 is reprinted in the Everyman edition.

master, had unfortunate ambitions in Logic and produced a theory of fallacies so weird and without antecedent or consequence that it is best passed over without further mention.[1] On the Continent, Gassendi (1592–1655) had been writing his history of Logic and made a collection of Megarian and Stoic examples of fallacy and paradox, but without systematization. (Some of these are also in Wilson's *Rule of Reason*.) The most influential logic text of the time was the *Hamburg Logic* (1638) of Joachim Junge (1587–1656) which, perhaps, deserves notice for the discovery, after almost two thousand years, of a way of classifying Aristotle's Fallacies Outside Language. The result is shown in the figure. The disjunction 'Common: Proper' had occurred earlier, as we saw, in Abraham Fraunce, and the 'forms of argument' referred to are Agricolan or Ramist Topics; but, for Fraunce, sophisms 'common to all places' had comprised only *Petitio principii* and Superfluitie. The disjunction 'inhering in antecedent: inhering in consequent' is also not quite new, having appeared in Buscher as a classification of 'faults of syllogism'; but, for Buscher, faults 'inhering in antecedent' comprise only those due to ambiguity of the middle term, and are not Fallacies Outside Language at all. Junge gives us no discussion of his classification, and, apart from this excursion into tree-building, his treatment is orthodox and rather scholastic.

Now let us move on to the *Port Royal Logic* of 1662, written by Antoine Arnauld, with or without the help of Pierre Nicole and other members of the Port Royal Movement in Paris. The correct name of this book is *The Art of Thinking*, itself a sign of a new approach to Logic, which had previously been regarded as teaching how to discuss, argue or reason, but never how to *think*. This is a very fine Logic book by any standards and is remarkably modern for its three centuries: it has influenced modern philosophy as much as any other book. Its strength is the closeness of the link it maintains with philosophical argumentation outside Logic, and particularly in its treatment of epistemological issues, following Descartes. Both Locke and Hume owe much to it. Yet its modernity is two-edged. In its discussion of 'the different ways of reasoning incorrectly' we find the origin

[1] Hobbes, *Computatio Sive Logica*, ch. 5. But see Engel, 'Hobbes's "Table of Absurdity".'

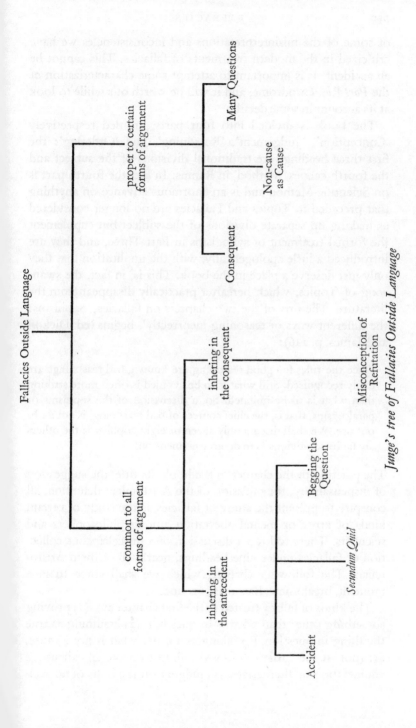

Junge's tree of Fallacies Outside Language

of some of the misinterpretations and inconsistencies we have criticized in the modern treatments of fallacies. This cannot be an accident. It is important to attempt some characterization of the *Port Royal* syndrome, and it will be worth our while to look at its account in some detail.

The book is divided into four parts, entitled respectively 'Conception', 'Judgement', 'Reasoning', and 'Ordering': the first three headings are traditional divisions of the subject and the fourth echoes 'Method' in Ramus. In fact the fourth part is on Scientific Method and is an enormous advance on anything that preceded it. Topics and Fallacies are no longer considered as making up separate divisions of the subject but supplement the formal treatment of syllogisms in Part Three, and they are introduced a little apologetically, with the implication that they only just deserve a place in the book. This is, in fact, the swansong of Topics, which hereafter practically disappear from the literature. The first of the two chapters on fallacies, 'Sophisms: the different ways of reasoning incorrectly', begins (ed. Dickoff and James, p. 246):

> Once the rules for good reasoning are known, bad reasonings are easily recognised. Still what is to be avoided is often more striking than what is to be imitated. So, a discussion of the sophisms or paralogisms, that is, the chief sources of bad reasoning, is not without use. We shall discuss only seven or eight sophisms; the others are faults so obvious as to deserve no mention.

The position of the chapter in the book, its title, the suggestion of dispensability, the omission of the Aristotelian definition, all conspire to present the study of fallacies as the study of certain kinds of error or mental aberration among philosophers and scientists. There follows a discussion of a miscellaneous collection of fallacies under nine headings, nearly all of them Aristotelian. The following chapter, which we shall come to in a moment, breaks new, Baconian, ground.

The kinds of fallacy treated in the first chapter are: (1) proving something other than what is in question, (2) assuming as true the thing in question, (3) taking as a cause what is not a cause, (4) (not in the earliest editions) failing to make an exhaustive enumeration of alternatives, (5) judging on the basis of an accidental characteristic, (6) passing from a divided to a connected sense or vice versa, (7) passing from a qualified to an absolute truth, (8) misusing the ambiguity of words, and (9) drawing a general conclusion from an incomplete induction. Only (4) and (9) are not, at least nominally, in Aristotle's list, and (4) could conceivably have been inspired by the fourth category of the Stoics;[1] (9) is clearly a fully-considered addition.

The first heading is linked with Aristotle's 'Misconception of Refutation', and alleged examples of it are drawn from Aristotle's own works. For the second (pp. 247–8):

> 2. The second sophism is to assume as true the very thing in question. This sophism Aristotle called begging the question (*petitio principii*). Since what serves as proof must be clearer and better known than what we seek to prove, we see easily enough that begging the question is altogether opposed to genuine reasoning. Galileo, however, has justly accused Aristotle of begging the question in his proof that the earth was at the centre of the universe.
>
> Aristotle argued:
>> The nature of heavy things is to tend toward the centre of the universe and of light things to tend away from the centre of the universe.
>>
>> Experience shows that heavy things tend toward the centre of the earth and that light things tend away from the centre of the earth.
>>
>> Therefore, the centre of the earth is the centre of the universe.
>
> The major premiss of this argument contains an obvious begging of the question. We see readily enough that what the second premiss states is true; but unless Aristotle assumed that the centre of the earth is the same as the centre of the universe – the very conclusion that he wished to prove by the argument – how did he learn that the major premiss is true?

This is only partly just: Arnauld's real quarrel is with Aristotle's aprioristic reasoning about 'the nature of heavy things' and Aristotle cannot clearly be accused of begging the question except on the assumption that his first premiss is intended to be empirical. The Fallacy of *Petitio Principii* is being invoked in an epistemological dispute. The impression is also given that the Fallacy resides in a syllogistic inference (though the syllogism in

[1] See above, p. 92.

question, on its own, is clearly valid) and we are well on the way to J. S. Mill's claim that every syllogism commits it. Arnauld gives three other examples of a similar kind.

Next we have (p. 250):

> 3. To take as a cause what is not a cause is the sophism called *non causa pro causa* and is a very common source of error. There are several ways to commit this sophism. One way is by simple ignorance of the true cause of a thing. For example, philosophers have credited any number of effects to nature's abhorrence of a vacuum; but some very recent and ingenious experiments, described in Pascal's excellent treatise, show that these effects are caused by the weight of air alone. These same philosophers commonly teach that a container filled with water breaks when the water freezes because the water contracts in freezing and leaves a vacuum which nature cannot endure. But quite the contrary has been discovered. The container breaks when the water freezes because water in freezing comes to occupy more space.

Arnauld was not the first to interpret 'taking as a cause what is not a cause' merely as 'giving an incorrect explanation', and we have seen that there is a hint of this interpretation in Aristotle's *Rhetoric*, despite its complete inconsistency with the account given in the *Sophistical Refutations* and elsewhere. Most of the Aristotelians, however, had been more or less faithful to the account in the *Sophistical Refutations* and none had departed from it so radically and completely as Arnauld. Under the new interpretation the fallacy is not 'logical', because it is not connected with any particular reasoning-process, valid or invalid. In subsequent examples apriorism, superstition, gullibility and pure obscurantism are the villains behind the false causal explanations, besides the Aristotelian *post hoc ergo propter hoc*, argument from temporal succession to causal influence (p. 255):

> ... people have concluded that the Dog Star is the cause of that extraordinary heat we feel during the dog days... But Gassendi has very correctly observed that nothing could be less likely than crediting the Dog Star with the heat of August. The Dog Star's influence ought to be strongest in the region to which the star is closest. But in August the Dog Star is much closer to the region below the equator than to us; and yet while we are in the dog days, the regions below the equator have their winter season.

But this example, like some modern ones, far from being a counter-example to *post hoc ergo propter hoc*, almost commits it itself, since one is led to assume that if it were *not* the case that the southern hemisphere is cold in August the causal explanation might escape unscathed. In this case we are up against a deep empiricist theoretical difficulty, the problem of induction.

The next heading but one reproduces a version of this difficulty (p. 259):

> 5. We commit a fifth kind of sophism when we make an unqualified judgment of a thing on the basis of an accidental characteristic. This sophism is called *fallacia accidentis* by the Schoolmen. For example, people commit this fallacy when they deprecate the use of antimony on the ground that when misused antimony produces bad effects. This fallacy is also committed by those who attribute to eloquence all the ill effects it works when abused or to medicine all the faults of ignorant doctors.

This hardly differs from 'taking as a cause what is not a cause'; but that it started a presently fashionable misinterpretation can be seen by comparing the examples with those of Accident in chapter 1 above.[1] A further major overlap occurs when we come to number (7), 'passing from a qualified truth to an absolute truth' because this, *a dicto secundum quid ad dictum simpliciter*, is quoted the other way round from the usual *a dicto simpliciter ad dictum secundum quid*. Since (9) gives us explicitly the case of incomplete induction, four of Arnauld's nine varieties nearly coincide.

Ambiguity is dealt with rather sketchily under (8), and it remains to say that Arnauld initiates another conceptual confusion by stating baldly 'All syllogisms invalid because of containing four terms are arguments which commit this fallacy'; without, that is, distinguishing use of a given term twice in different senses from use of two, possibly both unambiguous, terms.

The broad causes of these misconceptions have already been indicated and the faults of the *Port Royal* are no more than a working-out of trends already in existence. We have not yet finished, however, because the second chapter on fallacies, entitled 'Concerning faulty arguments advanced in public life and everyday affairs', represents a development in a different

[1] pp. 27-8.

direction and adds a further element to the typical modern account. Arnauld seems to have realized that his previous chapter is narrow in its range of application and, in a passage rewritten for the second edition, he says (pp. 265-6):

> But reason does not find its principal use in science; to err in science is not a grave matter, for science has but little bearing on the conduct of life...

There follows an attempt to distinguish 'false judgment' from 'bad reasoning'.

> Our errors can be credited to two principal sources: The one source – the will's dissoluteness – which troubles and perverts the judgment is internal; the other source is external and lies in the objects of which we judge and in their being able to deceive our minds by false appearances.

The mention of 'false appearances' echoes Bacon and the passage sounds like an attempt to separate two concepts conflated in the Idols. Arnauld affects to discuss the two sources of error separately but the bulk of the chapter is concerned with the way in which the emotions pervert reason.

Since this is a recurrent theme in future treatments of fallacies, something should be said about its genesis and history. For Plato, in Book IV of the *Republic*, the mind has three parts, intellect, spirit, and desire. He argues, first, that we observe three kinds of 'virtue' in the community – love of learning, spirit, and love of riches – and that these must be reflections of the same three kinds in the individual. The spirited character of Thracians and Scythians, he says, must be due to a spirited element in the minds of individual citizens; Athenians' love of learning must be due to domination of their minds by an intellectual element; and the love of riches of the commercially-minded Phoenicians and Egyptians by a desiring element. The three elements, however, must co-exist in a single mind just as the three corresponding classes, government, military, and commercial, must co-exist in society. A second argument for the existence of the three parts is that the mind is sometimes divided against itself: when we are driven by hunger and thirst intellect sometimes intervenes and we decline to eat and drink; the spirited element, in the case of the soldier, in battle may fight against the desires, and so on.

Aristotle, in a brief reference, criticizes Plato for thinking that the mind must have different *parts* in order to have different *functions* (*The Soul*, 432a 25); and his account of the mind's functions, which dominated Psychology throughout the Middle Ages, is a wider-ranging one which does not attempt to arrange them in neat categories. Since the seventeenth century, however, a distinction along Platonic lines has been revived as a kind of official starting-point for the study of mind. The distinction between 'action' and 'passion' is perhaps central; and there is no doubting, in Descartes, the completeness of the cleavage between the two parts or functions, nor the fact that the former, like Plato's guardians, is regarded as properly dominant. This appears not only in his *Passions of the Soul* but in, for example, his letters to Princess Elisabeth of Bohemia, where she seeks advice in the control of her troublesome passions and is duly instructed in their proper place in the scheme of things.[1] For Descartes, the important thing is that reason and will, loosely identified, should be dominant: for Hume (1711–76) later, the will was a 'direct passion' and also needed to be kept under control. Writing on 'Unphilosophical Probability', Hume considers (*Treatise*, vol. 1, p. 148):

> the case of a man, who being hung out from a high tower in a cage of iron cannot forbear trembling when he surveys the precipice below him, tho' he knows himself to be perfectly secure from falling, by his experience of the solidity of the iron, which supports him; ... The circumstances of depth and descent strike so strongly upon him, that their influence cannot be destroy'd by the contrary circumstances of support and solidity, ...

Here even the will is not to be trusted as much as cool reason; or, at least, it must be enlisted on reason's side.

But let us return to Arnauld and the *Port Royal*. The second chapter on Fallacies starts by considering 'Sophisms of Self-love, of Interest, and of Passion' and Arnauld is scathing about the stupidity of people who go in for arguments which he takes to be of the forms (pp. 266–8):

---

[1] Descartes, *Correspondance*, vol. 6; see particularly letter of 1 September 1645. On the development of this attitude see Levi, *French Moralists, the Theory of the Passions, 1585–1649*.

> I am a native of country X.
> Therefore, I must believe that Saint so-and-so preached in country X.
>
> I belong to the Y order.
> Therefore, I must believe that order Y has such-and-such privileges.

His formulation is unfortunate: he is not interested in the (rare) cases in which these arguments might be explicit so much as in cases in which analysis of the reasoner's attitude indicates that self-interest (or self-love, or passion) is the predominant causal factor. In many of these cases the reasoner will not be conscious of his motivation and will certainly not make it explicit. Incidentally, Bacon's Idols of the (Nation or) Tribe seem to be in evidence, though the examples do not represent precisely what Bacon had in mind. Arnauld goes on in the same vein to

> I like him.
> Therefore, he is the cleverest man in the world.
>
> I hate him.
> Therefore, he is a nobody.

and to

> If this were the case, I should not be a clever man.
> But, I am a clever man.
> Therefore, this is not the case.

and others. The only differences of any importance between the examples is in the nature of the considerations of interest or emotion that are supposed to be operative.

When 'Faulty Arguments Arising from the Objects Themselves' are considered, the same kind of analysis is, in effect, offered. The chapter finishes with a description of 'sophisms of authority' and 'sophisms of the manner'. Under the first heading we find the first appearance – unnamed – of the *argumentum ad baculum*:[1]

> The very manner in which some religious tenets are urged on us determines their credibility. In different ages of the Church – principally in the last century – we have seen men trying to spread their

---

[1] p. 287. See also, however, Wilson, *Rule of Reason*, f. 166: '... they fell to reasonyng with Argumentes, that ware neyther in fygure, nor in mode, but stode in plaine buffettes, whiche is a subtiltie, that is not mentioned within the compasse of this boke, ...'

doctrines by sword and bloodshed; we have seen men arm themselves against the Church by schism, against temporal powers by revolt; we have seen men without a common mission and without miracles and without any external marks of piety but rather with the obvious marks of licentiousness undertake to change the faith and the discipline of the Church. Any reasonable person will reject whatever is urged in so offensive a manner and not even the most stupid will listen.

Force, we should notice, is regarded not merely as irrelevant to an argument's merits but as positively injurious to them. Much the same appears to be true of testimony, at least when many people are involved:

Men follow the ridiculous procedure of believing a thing true according to the number of witnesses to its truth. A contemporary author has wisely pointed out that in difficult matters that are left to the province of reason, it is more likely that an individual will discover the truth than that many will.

Even discussion and debate may not help us here (p. 274):

Helpful as debates are when rightly used and when not invaded by passion, yet they are dangerous when improperly used by persons who pride themselves on maintaining their own opinions at any cost and on contradicting all other opinions.

On the other hand, although 'if any error is pardonable, it is the error of excessive deference to the opinion of a good man', people all too commonly reason in the following ways (p. 288):

He has an income of a hundred thousand a year.
Therefore, his judgment is good.

He is of high birth.
Therefore, what he advances is true.

He is a man who owns nothing.
Therefore, he is wrong.

With these examples of (under the modern misinterpretation of that term) *argumentum ad hominem*, we may begin to wonder how anyone ever succeeds in reasoning correctly. And the truth is that Arnauld does not give us any firm criteria, anywhere in this chapter, for distinguishing the not-so-bad extrinsic arguments from the plainly vicious. In this failing, again, he was ahead of his time.

Setting Arnauld against Bacon, we could find it hard to convince ourselves that they are from the same century. They both search for psychological classifications where previous writers had looked mainly to logical ones; but Bacon strikes us as no more than a step on our historical road whereas, with the *Port Royal*, we have an account that could nearly pass as contemporary. It is a pity that this is not an unqualified advantage. It was, however, the first and last such book of its time. The works from England in this period are not noted for originality. The *Institutio Logicae* (1686) of the mathematician John Wallis (1616-1703) is a competent piece of work which yet in no way matches the *Port Royal*. The better-known *Compendium* (1691) of Aldrich (1647-1710) draws heavily on Wallis but omits the best of him and displays a new rigidity like that of late scholasticism. Both Wallis and Aldrich write in Latin. The experiment of dropping Fallacies and keeping Topics having failed, Aldrich was to try the contrary one of dropping Topics and keeping Fallacies; and this, as we know, has been a success. For the rest, this is a dull little book, an affair of bare bones, and its author is more worthy of memory for his contributions to the architecture of Oxford, namely, Peckwater Quad and All Saints' Church.

We have reached, in fact, a period of logical decline. To some extent this is associated with continuing anti-Aristotelianism, which is as open in John Locke as in most of the Ramists: the arguments are the same. Ramus had said that Dialectic is natural to the human spirit and does not need to be taught: Locke says[1]

> ... God has not been so sparing to men to make them barely two-legged creatures, and left it to Aristotle to make them rational.

Locke's strictures against *maxims* in the early part of his Essay are in reference to the prevailing Topic-logic which really derives from Agricola and Ramus, though he does not dissociate it from 'scholastic Aristotelianism'. Thus (Book 1, ch. 2, § 27):

> ... But he that from a child untaught, or a wild inhabitant of the woods, will expect these abstract maxims and reputed principles of sciences, will, I fear, find himself mistaken ... They are the language and business of the schools and academies of learned nations,

[1] Locke (1632-1704), *Essay*, Book 4, ch. 17.

accustomed to that sort of conversation or learning where disputes are frequent; these maxims being suited to artificial argumentation and useful for conviction; but not much conducing to the discovery of truth or advancement of knowledge.

Although Locke's *Essay* was not published till 1690 there is evidence that he started writing it long before; probably about 1671, less than ten years after the first appearance of the *Port Royal*. He has read the *Port Royal*'s chapters on fallacies and gives this subject quite a lot of attention but integrates his observations more fully with his other work than, perhaps, any writer before or since. Thus he has several short chapters on the imperfections of Ideas – clear and obscure, distinct and confused (cf. Descartes), real and fantastical (cf. Hobbes), adequate and inadequate – and also, in Book III, 'Of Words', chapters entitled 'Of the Imperfections of Words', 'Of the Abuse of Words' and 'Of the Remedies of the foregoing Imperfections and Abuses', the last a most desirable, if not in fact very well-executed, addition. In Book IV, 'Of Knowledge and Opinion', there is a chapter 'Of Wrong Assent, or Error', and there are various discussions in other chapters that are generally relevant. Yet all this excellent work is largely unremembered compared with one short passage near the end of the chapter 'Of Reason'. The passage is parenthetical to the main discussion and has usually been omitted without ceremony by those who have prepared abridged editions.[1] It is worth quoting in full:

19. Before we quit this subject, it may be worth our while a little to reflect on *four sorts of arguments* that men, in their reasonings with others, do ordinarily make use of to prevail on their assent, or at least so to awe them as to silence their opposition.

First, The first is to allege the opinions of men whose parts, learning, eminency, power, or some other cause has gained a name and settled their reputation in the common esteem with some kind of authority. When men are established in any kind of dignity, it is thought a breach of modesty for others to derogate any way from it, and question the authority of men who are in possession of it. This is apt to be censured as carrying with it too much of pride,

---

[1] For example, it is not in the Pringle-Pattison edition, Oxford, 1924; or in the older Everyman Library edition, edited by Raymond Wilburn, London, 1947. It is here quoted from the Yolton edition, London, 1961; vol. 2, pp. 278–9.

when a man does not readily yield to the determination of approved authors which is wont to be received with respect and submission by others; and it is looked upon as insolence for a man to set up and adhere to his own opinion against the current stream of antiquity, or to put it in the balance against that of some learned doctor or otherwise approved writer. Whoever backs his tenets with such authorities thinks he ought thereby to carry the cause, and is ready to style it impudence in anyone who shall stand out against them. This I think may be called *argumentum ad verecundiam*.

20. *Secondly*, Another way that men ordinarily use to drive others and force them to submit their judgments and receive the opinion in debate is to require the adversary to admit what they allege as a proof, or to assign a better. And this I call *argumentum ad ignorantiam*.

21. *Thirdly*, A third way is to press a man with consequences drawn from his own principles or concessions. This is already known under the name of *argumentum ad hominem*.

22. *Fourthly*, The fourth is the using of proofs drawn from any of the foundations of knowledge or probability. This I call *argumentum ad judicium*. This alone of all the four brings true instruction with it and advances us in our way to knowledge. For: (1) It argues not another man's opinion to be right because I, out of respect or any other consideration but that of conviction, will not contradict him. (2) It proves not another man to be in the right way, nor that I ought to take the same with him, because I know not a better. (3) Nor does it follow that another man is in the right way because he has shown me that I am in the wrong. I may be modest and therefore not oppose another man's persuasion; I may be ignorant and not be able to produce a better; I may be in an error and another may show me that I am so. This may dispose me, perhaps, for the reception of truth but helps me not to it; that must come from proofs and arguments and light arising from the nature of things themselves, and not from my shamefacedness, ignorance, or error.

That is all: the passage is self-contained. This is not intended as a classificatory treatment of fallacies, least of all in the *Port Royal* sense, and he does not explicitly link these few remarks with what he says about Error, or the Abuse of Words. What might strike us about this passage, in fact, is its Aristotelian tone: this is a classification of assent-producing devices, sophistical or otherwise. We can push the Aristotelian analogy a little further than this; for the *argumentum ad judicium* is surely close, in description and honorific intent, to Aristotle's 'didactic' or 'demonstrative' arguments, 'those that reason from principles appropriate to each subject'; and the *argumentum ad hominem*, in which we 'press a man with consequence drawn from his own principles or concessions', resembles Aristotle's 'dialectical' arguments, at least as they are interpreted in the context of Greek debate; and the others are 'contentious'. What is even more Aristotelian about the passage is that Locke does not clearly condemn any of the argument-types, but stands poised between acceptance and disapproval. It is clear from the final paragraph that arguments of the first three types are less than perfect, but not clear that they are never to be used; at least, in practical politics.

Locke indicates clearly that he has invented three of the terms himself, but that the *argumentum ad hominem* is already known under that name. This poses the interesting question: where did he get it from? We may search the writings of Locke's contemporaries and immediate predecessors for it in vain: it is not in Bacon, or Hobbes, or the *Port Royal*. We might have hoped to find it in legal tradition, since it would seem to have some relevance to possible abuses of Court-room examination; but it is not, for example, in Abraham Fraunce's *Lawiers Logike*. The Oxford Dictionary gives its use by Locke as the first occurrence in English, and Larousse gives Bossuet as introducing it into French about the same time. Where did it come from?

The answer – perhaps not surprisingly, in view of what we have just been saying – is that it comes from Aristotle. In the second half of the *Sophistical Refutations* Aristotle says of a proposed solution of one of the examples of the Fallacy of Division that it is 'valid against the questioner, but not against his argument' (177b 33); that is, that it is only against certain ways of putting the questions that the proposed solution is valid. A little later he makes the same kind of point in the case of another example, and says (178b 17; see also 183a 21):

> ... these persons direct their solutions against the man, not against his argument.

The vulgate Latin of this passage (Migne, *Patrologia Latina*, vol.

64, col. 1031) is . . . *hi omnes non ad orationem, sed ad hominem solvunt*, . . . and the distinction between *ad orationem* and *ad hominem* appears here and there in medieval commentaries and treatises.¹ Locke, however, may have had to go back a long way to find this term. The weaker and stronger senses of 'refutation', namely (*a*) destruction of an opponent's proof, and (*b*) construction of the proof of a contrary thesis, have seldom been clearly distinguished and we have already explored some of the difficulties in the concepts involved in the *dubitatio* version of the Obligation game.

Locke's *argumentum ad ignorantiam* touches on another feature of argumentation that is all too often regarded as no business of the logician; the question of burden of proof. What Locke says is a little less than adequate here, in that he apparently assumes that parties to an argument are always all equally under obligation to prove their respective cases; but again, is more tolerant towards the argument as a dialectical move than either the *Port Royal* or most modern authors.

Only the *argumentum ad verecundiam*, in fact, has anything of the modern flavour attached to it; though it is worth noticing that *verecundia* means 'modesty' and that Locke is explicit in referring to men's natural reluctance to challenge authorities that are learned, eminent or powerful. He is not referring to the fallacy of relying on *worthless* authorities so much as with the reminder that *even worthy* authorities, whom it is normally reasonable to trust, may be wrong. Hence we misrepresent even the *argumentum ad verecundiam* if we treat it in the manner of the *Port Royal*.

Leibniz (1646–1716), though he admired much in Aristotle, wrote nothing on fallacies except by way of commentary on Locke. Characteristically, he suggests (*New Essays*, Bk. 4, ch. 17, p. 577) as an addition to Locke's four *argumenta* the *argumentum ad vertiginem* (giddiness):

> . . . when we reason thus: if this proof is not received we have no means of attaining certainty upon the point in question, which we take as an absurdity. This argument is valid in certain cases, as if any one wished to deny primitive and immediate truths, for example, that anything can be and not be at the same time, or that we ourselves exist, . . .

¹ See the *Summa*, in De Rijk, *Logica Modernorum*, vol. 1, p. 430; and the commentary of Albert the Great on the *Sophistical Refutations*, Book 2, tract 2, chapters 3 and 6 and tract 5, chapter 1.

But on less basic matters arguments *ad vertiginem*, he says, are fallacious. Like so much in Leibniz's writings, this suggestion bore no fruit.

As an anti-Aristotelian, Locke was more effective than Ramus had been: he not only turned his readers against traditional Logic but also absorbed and digested so much of it in his own work that they did not feel the deprivation. He received support, on the Continent, from the developing rationalist movement, which drove Philosophy off in yet another direction. The result was that there was negligible new work of any kind on the subjects that concern us. The needs of elementary teaching were filled by books already in existence, such as Aldrich's *Compendium*.

The eighteenth century, despite giants like Hume and Kant, was another Dark Age for Logic; with only a few immature stirrings behind the scenes, from writers like Saccheri and Ploucquet, to give promise for the future. Almost the only logic book written in English during the entire century was one in 1725 by Issac Watts, better known as the author of 'O God, our help in ages past'[1] Watts draws on the *Port Royal* and has a lot to say about Prejudices, arising from Things, from Words, from Ourselves and from Other Persons, as well as about Sophisms in the traditional sense: his selection of the latter is the *Port Royal*'s selection, in the same order, and with only a few changes, mainly for the worse. He mentions in its own right Locke's classification of arguments, but in such a way as to give the eventual combination of this with the classification of fallacies a good push forward (*Logick*, pp. 465–6):

> There is yet another Rank of Arguments which have *Latin* Names; their true Distinction is derived from the Topics or middle Terms which are used in them, tho' they are called an Address to our *Judgment*, our *Faith*, our *Ignorance*, our *Profession*, our *Modesty*, and our *Passions*.
>
> 1. If an Argument be taken from the Nature or Existence of Things, and addrest to the *Reason* of *Mankind*, 'tis called *Argumentum ad Judicium*.

---

[1] There was one other such book, the *Elements of Logick* (1748) of William Duncan. 'Murray's Logic', originally in Latin, had also been used as a textbook; and there were books by John Wesley, Henry Home, and John Collard.

2. When 'tis borrowed from some convincing Testimony, 'tis *Argumentum ad Fidem*, an Address to our *Faith*.

3. When 'tis drawn from any insufficient *Medium* whatsoever, where the Opposer has not Skill to refute or answer it, this is *Argumentum ad Ignorantiam*, an Address to our *Ignorance*.

4. When 'tis built upon the profest Principles or Opinions of the Person with whom we argue, whether these Opinions be true or false, 'tis named *Argumentum ad Hominem*, an address to our *profest Principles*. *St. Paul* often uses this Argument when he reasons with the *Jews*, and when he says, *I Speak as a Man*.

5. When the Argument is fetch'd from the Sentiments of some wise, great, or good Men, whose Authority we reverence, and hardly dare oppose, 'tis called *Argumentum ad Verecundiam*, an Address to our *Modesty*.

6. I add finally, when an Argument is borrowed from any Topics which are suited to engage the Inclinations and Passions of the Hearers on the Side of the Speaker, rather than to convince the Judgment, this is *Argumentum ad Passiones*, an Address to the *Passions:* or if it be made publickly, 'tis called an *Appeal to the People*.

The *argumentum ad fidem* and the *argumentum ad passiones* (or *ad populum*) have been added to Locke's list; and the description of the others subtly altered in perversion of Locke's original intentions: possibly there was an oral or teaching tradition by this time that Watts was reproducing.

The later eighteenth century replaced Logic with Rhetoric, conceived, as in the sixteenth, as a science of literary style; for example, in the work of Campbell (1719–96). For the rest, the period has no interest for us but, before shutting the book on it we might just notice Laplace's (1749–1827) *Philosophical Essay on Probabilities* whose first draft dates from 1795, though it was not published till 1819. Besides trying to apply the theory of Probability to the problem of evaluating testimony, Laplace has a chapter 'Concerning Illusions in the Estimation of Probabilities'. One such illusion is what has become known as the 'gambler's fallacy' or fallacy of 'maturity of the chances', the illusion that a run of events of a certain kind makes a run of contrary events more likely in order to even up the proportions (p. 162):

I have seen men, ardently desirous of having a son, who could learn only with anxiety of the births of boys in the month when they expected to become fathers. Imagining that the ratio of these births to those of girls ought to be the same at the end of each month, they judged that the boys already born would render more probable the births next of girls.

This is an illusion generated by probability-theorists themselves. It is neatly cancelled by a contrary illusion (p. 163):

> ... one seeks in the past drawings of the lottery of France the numbers most often drawn, in order to form combinations upon which one thinks to place the stake to advantage.

This one has no special name, unless we allow '*im*maturity of the chances'. Laplace's is not the last word on these fallacies.

The project of a political *Book of Fallacies* was conceived by Jeremy Bentham as early as 1788, but it did not come to fruition till much later. Bentham had a large circle of friends in various depths of proselytism and first tried to persuade a young Frenchman called Étienne-Louis Dumont to ghost the book for him from rough notes and fragments. Dumont got caught up in the French Revolution but eventually finished a version of the book in French and it was published in 1816.[1] Whether Bentham was dissatisfied with Dumont's version we do not know; but at all events he made a move to have a new version written in English. This was finally published in 1824, and the hand that actually held the pen was that of a young lawyer named Peregrine Bingham. As 'editor' of the work Bingham seems to have had a fair freedom even in the matter of terminology and, in the introduction, describes his principles of classification. Bentham had wanted to divide the book into Fallacies of the *Ins*, Fallacies of the *Outs*, and *Either-side* Fallacies and had several other overlapping or supplementary suggestions; but Bingham 'preferred Dumont's arrangement to that pursued by the author'. The main division into Fallacies of Authority, Danger, Delay, and

---

[1] The *Traité des sophismes politiques* is the second part of the *Tactique des assemblées législatives* and was published under Bentham's name in Paris in 1816. See the introduction by Crane Brinton to Bentham, *Handbook of Political Fallacies*: this is a reprint, with some unacknowledged alterations by the editor, of the original *Book of Fallacies: from Unfinished Papers of Jeremy Bentham*, by A Friend, London, 1824.

Confusion is not of great importance: the individual headings are many and weird, 'The Hobgoblin Argument, or "No Innovation"', 'Official Malefactor's Screen, or "Attack Us, You Attack Government"', 'Fallacy of Artful Diversion', 'Opposer-General's Justification: Not Measures but Men, or, Not Men but Measures', and so on, and so on. Latin tags such as *ad verecundiam*, or *ad superstitionem*, or any one of a dozen others, are used as subtitles ('not for ostentation, but for prominence, impressiveness, and hence for clearness' says Bingham's introduction) and, oddly, they include *ad judicium*, which tags a number of the Fallacies of Confusion. A few extracts, necessarily rather out of context:

The wisdom of our ancestors: or Chinese argument.

Ad verecundiam.

This argument consists in stating a supposed repugnancy between the proposed measure and the opinions of men by whom the country of those who are discussing the measure was inhabited in former times; these opinions being collected either from the express words of some writer living at the period of time in question, or from laws or institutions that were then in existence... (*Book of Fallacies*, p. 69).

Procrastinator's Argument.

Ad socordiam.

'Wait a little, this is not the time'.

... This is the sort of argument or observation which we so often see employed by those who, being in wish and endeavour hostile to a measure, are afraid or ashamed of being seen to be so. They pretend, perhaps, to approve of the measure; they only differ as to the proper time of bringing it forward; but it may be matter of question whether, in any one instance, this observation was applied to a measure by a man whose wish it was not, that it should remain excluded for ever (pp. 198–9).

Question-begging Appellatives.

Ad judicium.

... Begging the question is one of the fallacies enumerated by Aristotle; but Aristotle has not pointed out (what it will be the object of this chapter to expose) the mode of using the fallacy with the greatest effect, and least risk of detection, – namely, by the employment of a single appellative...

... it neither requires nor so much as admits of being taught ... The great difficulty is to unlearn it: in the case of this, as of so many other fallacies, by teaching it, the humble endeavour here is to unteach it.

... Take, for example, *improvement* and *innovation:* under its own name to pass censure on any improvement might be too bold: applied to such an object, any expressions of censure you could employ might lose their force: employing them, you would seem to be running on in the track of self-contradiction and nonsense.

But improvement means something new, and so does *innovation*. Happily for your purpose, *innovation* has contracted a bad sense; it means something which is new and bad at the same time. Improvement, it is true, in indicating something new, indicates something good at the same time; ... (pp. 213-18).

Sham Distinctions.

Ad judicium.

... When any existing state of things has too much evil in it to be defensible *in toto*, ... declare your approbation of the good by its eulogistic name, and thus reserve to yourself the advantage of opposing it without reproach by its dyslogistic name, ...

Example 1. – *Liberty and Licentiousness of the Press* ...

Example 2. – *Reform, temperate and intemperate* ... (pp. 271-6).

Particular demand for Fallacies under the English Constitution.

Two considerations will suffice to render it apparent that, under the British Constitution, there cannot but exist, on the one hand, such a demand for fallacies, and, on the other hand, such a supply of them, as for copiousness and variety, taken together, cannot be to be matched elsewhere.

1. In the first place, a thing necessary to the existence of the demand is discussion to a certain degree free.

Where there are no such institutions as a popular assembly taking an efficient part in the Government, and publishing or suffering to be published accounts of its debates, – nor yet any free discussion through the medium of the press, – there is, consequently, no demand for fallacies. Fallacy is fraud, and fraud is useless when every thing may be done by force.

The only case which can enter into comparison with the English Government, is that of the United Anglo-American States ... (p. 389).

Finally, Bentham suggests that in the publication of parliamentary reports such as *Hansard* special marks of reference should be used to denote the occurrence of fallacies, and says (p. 410):

> In the course of time when these imperfect sketches shall have received perfection and polish from some more skilful hand...[when any legislator shall be] so far off his guard as through craft or simplicity to let drop any of these irrelevant, and at one time deceptious arguments ... instead of, Order! Order! a voice shall be heard, followed, if need be, by voices in scores, crying aloud, 'Stale! Stale! Fallacy of authority, Fallacy of distrust,' Etc. Etc.

Bentham is no Machiavelli: he always teaches in order to unteach.

The Rev. Sydney Smith, who reviewed the *Book of Fallacies* in the *Edinburgh Review* in 1825, decided that he could best illustrate them by writing a speech which systematically perpetrated all of them in turn. He called it 'The Noodle's Oration':

> What would our ancestors say to this, Sir? How does this measure tally with their institutions? ... Is beardless youth to show no respect for the decisions of mature age? (Loud cries of Hear! Hear!)...
>
> Besides, Sir, if the measure itself is good, I ask the honourable gentleman if this is the time for carrying it into execution– ... If this were an ordinary measure, I should not oppose it with so much vehemence; but, Sir, it calls in question the wisdom of an irrevocable law– ...
>
> ... but what is there behind? What are the honourable gentleman's future schemes? If we pass this bill, what fresh concessions may he not require? ...
>
> I profess myself, Sir, an honest and upright member of the British Parliament, and I am not afraid to profess myself an enemy to all change, and all innovation. I am satisfied with things as they are; ...

Not everyone, however, was as pleased with Bentham's efforts. Richard Whately, who had stronger feelings about Logic than Sydney Smith, and whom we must go on to consider in his own right, said[1]

---

[1] In the article on 'Rhetoric' in the *Encyclopaedia Metropolitana*, vol. 1, ch. II, p. 265, published about 1817. Whately wrote the articles on 'Logic' and 'Rhetoric' and later expanded both into book form: *Elements of Logic* (1826) and *Elements of Rhetoric* (1828). The passage quoted was, however, omitted from the *Elements of Rhetoric*.

It is matter of regret that the powers of such a mind as that of Mr. Bentham, should be to so great a degree wasted. Such, however, must always be the case, when a Scientific work is composed (with whatever sincerity) for party purposes, or with any object foreign to the precise End of the Science in question.

Bentham is accused of perpetually committing his own Fallacies: *petitio principii*, vituperative language and, in place of a blind veneration for authority, 'an equally blind craving after novelty for its own sake, and a veneration for the ingenuity of one's own inventions'.

Whately, who taught at Oxford and was later Archbishop of Dublin, understood the rhetorical tradition which preceded him but felt the need in it for the greater precision of Logic. We should not underestimate his efforts and success in renewing interest in Logic. At Oxford, Aldrich was the main text, but was too abridged and sketchy for any but elementary courses. Watts he found unsatisfactory, and he had to go back to Wallis for any improvement. His *Logic* was immediately successful both in Britain and the United States.[1] It was, perhaps, partly Bentham's doing that he felt the need to do something, in particular, about the study of fallacies. However, he was much more concerned than Bentham to get to grips with the traditional material. He says (*Elements of Logic*, Bk. III, Intro. pp. 101-2):

> It is on Logical principles therefore that I propose to discuss the subject of Fallacies; and it may, indeed, seem to have been unnecessary to make any apology for so doing, after what has been formerly said, generally, in defence of Logic; but that the generality of Logical writers have usually followed so opposite a plan. Whenever they have to treat of any thing that is beyond the mere elements of Logic, they totally lay aside all reference to the principles they have been occupied in establishing and explaining, and have recourse to a loose, vague and popular kind of language; . . .

This is a familiar modern complaint; and we begin to look forward to seeing Whately meet it.

The tree of classification that he evolved to replace the traditional one is reproduced on page 171. Its most significant feature is a reappraised allocation of fallacies into two categories,

---

[1] The *Metropolitana* article was used, 'I believe, in every one of their Colleges'; *Elements of Logic*, preface, p. x.

Logical and Non-Logical; and Aristotle's 2,000-year-old qualms about his own twofold division are at last set at rest. Whether Whately's classification precisely achieves Aristotle's aim is not important; but Whately at least provides a simple and clear criterion for his dichotomy (Bk. III, § 2):

> In every Fallacy, the Conclusion either *does*, or *does not follow from the Premises*. Where the Conclusion does not follow from the Premises, it is manifest that the fault is in the *Reasoning*, and in that alone; these, therefore, we call Logical Fallacies, as being properly, violations of those rules of Reasoning which it is the province of Logic to lay down.

These include not only those 'exhibiting their fallaciousness by the bare *form* of the expression, without any regard to the meaning of the terms', but also all cases of ambiguity; for although Logic cannot itself tell us *how to find* fallacies of this kind, it can tell us *where to search* for them. Non-Logical fallacies comprise mainly question-begging and various forms of irrelevant conclusion, in which the whole point is that a fallacy may be committed in spite of the presence of some kind of valid inference; and Aldrich is given a short blast for apparently complaining, in the passage quoted above,[1] that they are not genuine cases.

What we can best learn from Whately, however, is not that fallacies can and should be reclassified – which nearly every account since him has told us – but rather some facts about the place in practical argumentation of the arguments '*ad*'. What he says is not very fully worked out, and it was not influential, but we shall want to take it up later. We must turn first to his account, in the *Elements of Rhetoric* (Part I, ch. III, § 2), of Presumption and Burden of Proof. This has, at times, been misunderstood.

When one is engaged in making a case for a conclusion it is very important in practice to be clear on which side the *Presumption* lies, and to which belongs the *Burden of Proof*. These are accepted legal concepts:

> According to the most correct use of the term, a 'Presumption' in favour of any supposition, means, not (as has been sometimes erroneously imagined) a preponderance of probability in its favour, but,

[1] p. 49.

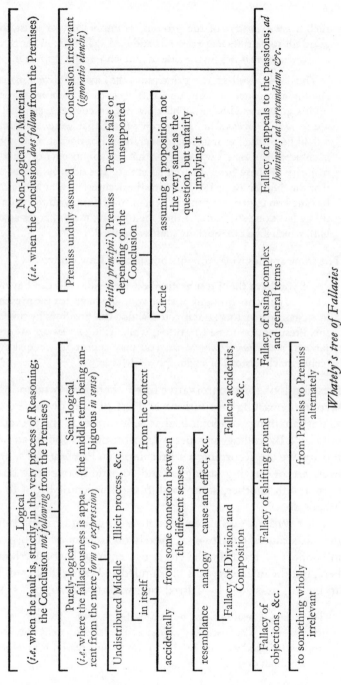

Whately's tree of Fallacies

such a *pre-occupation* of the ground, as implies that it must stand good till some sufficient reason is adduced against it; in short, that the *Burden of proof* lies on the side of him who would dispute it.

Thus, it is a well-known principle of the Law, that every man (including a prisoner brought up for trial) is to be *presumed* innocent till his guilt is established. This does not, of course, mean that we are to *take for granted* he is innocent; for if that were the case, he would be entitled to immediate liberation: nor does it mean that it is antecedently *more likely than not* that he is innocent; or, that the majority of these brought to trial are so. It evidently means only that the 'burden of proof' lies with the accusers; – that he is not to be called on to prove his innocence, or to be dealt with as a criminal till he has done so; but that they are to bring their charges against him, which if he can repel, he stands acquitted.

The same applies to arguments outside the Law Courts.

> ... if you have the 'Presumption' on your side, and can but *refute* all the arguments brought against you, you have, for the present at least, gained a victory: but if you abandon this position, by suffering this Presumption to be forgotten, which is in fact *leaving out one of, perhaps, your strongest arguments*, you may appear to be making a feeble attack, instead of a triumphant defense.

Presumptions are a conservative force: there is a Presumption in favour of existing institutions and established doctrines, and against anything paradoxical, that is, 'contrary to the prevailing opinion'. By calling any person, institution or book an 'Authority' we are according it a Presumption. Presumptions, however, are of varying strengths and sometimes one may be rebutted with another, so as to shift the Burden of Proof to the other side.

What Whately says about the conservative nature of presumptions has caused controversy, largely on account of his use of it in support of the Church. He is not entirely beyond criticism here, but the main charge is ill-founded and easily answered. His argument, as stated by Alfred Sidgwick (*Fallacies*, p. 155), is

> 'There is a Presumption' he writes 'in favour of every *existing* institution' ... 'Christianity *exists*; and those who deny the divine origin attributed to it are bound to show some reason for assigning it to a human origin.'

I have reproduced Sidgwick's punctuation exactly in this passage; but the row of dots in the middle of the argument marks the omission of about two pages of Whately's intervening discussion. In fact the passage starts:

> There is a Presumption in favour of every *existing* institution. Many of these (we will suppose, the majority) may be susceptible of alteration for the better; but still the 'Burden of proof' lies with him who proposes an alteration; simply, on the ground that since a change is not a good in itself, he who demands a change should show cause for it. No one is *called on* (though he may find it advisable) to defend an existing institution, till some argument is adduced against it; and that argument ought in fairness to prove, not merely an actual inconvenience, but the possibility of a change for the better.

Whately fails, perhaps, to distinguish clearly between an institution as such and a set of institutionalized beliefs, and he does not explore at all the question of how entrenched an institution needs to be in order to command adherence prima facie. He is quite clear, however, that the presumption he speaks of is not itself the kind of thing that can carry weight *as an argument*: it merely decides which party, if there is to be an argument, must lead the attack. Apologist for Christianity he undoubtedly was, but he was not – as Sidgwick's presentation makes him appear – so cavalier as to rest an argument for it, *ad verecundiam*, on the existence of the Church of England.

The attempt to shift the burden of proof to one's opponent is a sophistical trick that Aristotle – since in the Greek game it always resides with the questioner – does not mention, and it seems appropriate that it should be considered in connection with a list of fallacies. But this is not all: it has a particularly relevant function when we come to consider extrinsic arguments, or arguments '*ad*'.

By Whately's time, the *Port Royal* account of the passions has become a dogma, and the marriage of the terms *ad hominem*, *ad verecundiam* and the others to this account has been consummated. Whately finds a place for them, with reservations, under the heading 'irrelevant conclusion', or *ignoratio elenchi* (Bk. III, §15).

There are certain kinds of argument recounted and named by Logical writers, which we should by no means universally call Fallacies; but which *when unfairly* used, and *so far as they are* fallacious, may very well be referred to the present head; such as the '*argumentum ad hominem*,' ['or personal argument,'] '*argumentum ad verecundiam*,' '*argumentum ad populum*,' etc. all of them regarded as contradistinguished from '*argumentum ad rem*,' or, according to others (meaning probably the very same thing) '*ad judicium*.' These have all been described in the lax and popular language before alluded to, but not scientifically: the '*argumentum ad hominem*,' they say, 'is addressed to the peculiar circumstances, character, avowed opinions, or past conduct of the individual, and therefore has a reference to him only, and does not bear directly and absolutely on the real question, as the "*argumentum ad rem*" does:' in like manner, the '*argumentum ad verecundiam*' is described as an appeal to our reverence for some respected authority, some venerable institution, etc. and the '*argumentum ad populum*,' as an appeal to the prejudices, passions, etc. of the multitude; and so of the rest. . .

It appears then (to speak rather more technically) that in the '*argumentum ad hominem*' the conclusion which actually is established, is not the *absolute* and *general* one in question, but *relative* and particular; *viz.* not that 'such and such is the fact,' but that '*this man* is bound to admit it, in conformity to his principles of Reasoning, or in consistency with his own conduct, situation,' Etc.

Here Whately adds a footnote:

The 'argumentum ad hominem' will often have the effect of shifting the *burden of proof*, not unjustly, to the adversary.

We are referred to the account of Burden of Proof in the *Elements of Rhetoric* and given the following example: When a sportsman is accused of barbarity in sacrificing hares or trout he may safely turn the tables by replying 'Why do you feed on the flesh of the harmless sheep and ox?' Whately continues, in the text:

Such a conclusion it is often both allowable and necessary to establish, in order to silence those who will not yield to fair general argument; or to convince those whose weakness and prejudices would not allow them to assign to it its due weight . . . provided it be done plainly, and *avowedly*; . . . The fallaciousness depends upon the *deceit*, or attempt to deceive.

The same analysis is to apply to the *argumentum ad verecundiam*, and the others.

The footnoted suggestion that arguments of these special kinds have a role in connection with the allocation of burden of proof – that they create presumptions – is an excellent and interesting one. It needs to be worked out in more detail, with some qualifications attached to it. It works excellently in accommodating *ad verecundiam*, reasonably well for some, but not all, cases of *ad hominem*, not at all so well for *ad populum*: it is not clear what other cases Whately has in mind. (All of these terms are here subject to their modern, historically inaccurate, interpretation. I have not been able to discover definitely where this interpretation originated, or what is the source of the quoted or pseudo-quoted description of *ad hominem*.) We might hope to improve on this account at least to the extent of sifting details; and could hope that someone, in the later nineteenth century, might have done so for us already.

This hope is not realized. Our history has been brought almost as nearly up to date as it is profitable to bring it. We may finish with brief notes on four other writers of the nineteenth century: Artur Schopenhauer, John Stuart Mill, Augustus De Morgan, and Alfred Sidgwick. None of these added anything very new to the study but all four provided original points of view.

Schopenhauer's *Art of Controversy* is the least historically-blinkered of any account of the subject. It starts with a discussion of the ancient concepts of Dialectic, Rhetoric, Eristic, and Sophistic, mainly in Aristotle, and attempts to delineate Dialectic in particular as *the art of getting the best of it in a dispute*. A rudimentary attempt is made to describe the 'basis of all Dialectic'. The rest of the essay is then a description of thirty-eight 'stratagems' of argument, some of which are from Aristotle, some perhaps from Bacon and some reminiscent of Bentham; many clearly original. The essay was not published complete until after Schopenhauer's death and cannot have had much influence on logical tradition.

John Stuart Mill had spent some part of his youth in ghostwriting for Bentham but, when he came to write his *System of Logic*, did not take over his mentor's rhetorical-political conception of argument. His examples, instead, are almost all from science. From Bentham and Whately, however, he inherited the urge to reclassify the traditional material. His main categories were enumerated above in chapter 1 and exhibit his concern with empirical investigation, but the content is less original than it appears.

Augustus De Morgan (1806–71), whose *Formal Logic* was published in the same year (1847) as George Boole's *Mathematical Analysis of Logic*, was a mathematician who was to contribute considerably to the modern logical movement that grew out of Boole's thought. In his discussion of fallacies in this earlier work he is often acute. He has read Aristotle and some unidentified scholastic writings but this does not stop him from misrepresenting False Cause and Accent: in the latter case he may fairly be held responsible for the modern travesty. Copi's example of a sentence that changes its meaning with change of emphasis, 'We should not speak ill of our friends', compares with De Morgan's 'Thou shalt not bear false witness against thy neighbour'. He invented others of the modern stock examples.

Alfred Sidgwick (1850–1943), cousin of the better-known Henry, is, so far as I can discover, the only person ever to have tried to develop a complete theory of Logic around the study of fallacies: 'Logic may in fact be viewed as a machine for combating Fallacy' (*Fallacies*, p. 11). The result is not, in this case, a success and has been passed over and left behind by modern developments. This does not mean that the project itself does not deserve attention. Sidgwick's priorities lead him to say a good deal about neglected topics like Presumption or Burden of Proof and (Mill's influence is evident) argument by Sign and Analogy. The latter have their characteristic risks and sources of error.

CHAPTER 5

The Indian Tradition

Despite large and obvious differences, the history of Indian Logic runs curiously parallel with that of the Logic of Europe, at least in its classical and medieval periods; so much so that one is tempted to see the two as advancing side by side, rather than separately. So far as the classical period is concerned we may, in fact, have tended to underestimate the extent of the contact between India and Greece; for though no one now maintains, as some historians have argued,[1] that Indian Logic was directly inspired by Aristotle, or even that influences in either direction can be definitely traced, it is known that there was considerable commerce between the two regions. Aristotle's most famous pupil Alexander had, after all, penetrated to India in the course of his military ventures, though it is doubtful that he found time to lecture the inhabitants on Logic. There were subsequently Greek settlements on the north-west frontier, and it would be surprising if there were not some interpenetration of ideas. Later, with the rise of Mohammedanism, these ties were broken, and India and Europe were held apart by a civilization largely hostile to both. The Arabs, it is true, helped to preserve the European tradition to some extent by taking over Aristotle's works in translation; but they did not themselves discover the Indian Logic and, in fact, the Moslem invasion of India nearly caused its extinction. Many of our modern Sanskrit texts are re-translations from versions that survived in Tibet. We can also study such doctrines as went with Buddhism to China.

The actual authors of the earliest Indian treatises are dim,

[1] Vidyābhūṣaṇa, *A History of Indian Logic*, Appendix B.

shadowy figures and it is difficult to date their works even approximately. The most important text, the *Nyāya sūtra* (to a sufficient approximation the name means just 'Logic Book'), was supposedly written by one Gautama or Akśapeda (two persons or one?) at some time during the first three centuries A.D.; but it could be a compilation, and one part of it, the fifth book, which will occupy us later, has some parallel in the work of the physician Caraka, who is placed about A.D. 70. It is quite short – about 12,000 words altogether in English translation – and aphoristic in style, to the extent that it needs to be accompanied by a detailed commentary. A commentary by Vātsyāyana (5th–6th century?), author also of the *Kama sūtra*, is very often printed with it.[1]

Before considering what Gautama and Vātsyāyana say about Fallacies it is necessary to sketch their theory of inference. This betrays a 'dialectical' origin almost as strongly as Aristotle's *Topics*. A single pattern of inference is given, without moods or figures. An inference has five members. Illustrated by the stock example of later writers, they are:

(1) (Thesis): 'The hill is fiery.'
(2) (Reason): 'Because it has smoke.'
(3) (Example): 'Whatever is smoky is fiery, like a kitchen.'
(4) (Application): 'And this hill is smoky.'
(5) (Conclusion): 'Therefore it is fiery.'

Since Gautama calls (3) just 'Example', it seems likely that, in the illustration, the statement of a general rule, 'Whatever is smoky is fiery', is a later importation: the original idea is simply that an example of the operation of the (unstated) connection between the major term 'fire' and middle term 'smoke' should be given. If this seems strange, comparison with Aristotle *Rhetoric* (1393a 22–1394a 18) might be helpful. In general the aim of this theory of inference is much closer to Aristotle's aim in his *Rhetoric* than it is to what we are used to and which derives from the *Prior Analytics*. Gautama defines an 'example' as a 'case in which the

---

[1] I have used mainly *Gautama's Nyāyasūtras*, in the Jhā edition, which also contains Vātsyāyana's commentary; and have supplemented it with some of the extracts given in Bocheński, *A History of Formal Logic*. I have altered terms here and there where it was necessary to bring the translations into correspondence; on the whole preferring Jhā.

common man and the expert agree': in this case there is some resemblance to Aristotle's definition of dialectical or examination arguments.[1]

The apparently useless repetitive character of (4) and (5) led to their being dropped by some later writers. Their presence is less surprising, however, if the whole inference scheme is seen as a *pro forma* for the setting-out of inferences in practice, with an aim of securing comprehension in an audience or persuasion of an opponent. Here it is interesting that we are told by Vātsyāyana in his commentary that others raised the number of members of the syllogism to ten, by adding the 'desire to know', 'doubt', 'belief in possibility of solution', 'purpose in view in attaining the conclusion' and 'removal of doubt'. If these are interleaved with the other five we have the pattern of a veritable dialogue, in which the added members represent the reactions or contributions of a second participant. We could dramatize the situation something as follows:

A: The hill is fiery. (*Thesis*)
B: Why? (*Desire to know*)
A: Because it is smoky. (*Reason*)
B: Does that follow? (*Doubt*)
A: As in the case of a kitchen. (*Example*)
B: Oh, I begin to see! (*Belief in possibility of a solution*)
A: And, you see, this hill is smoky. (*Application*)
B: Now we are getting somewhere. (*Purpose in view in attaining the conclusion*)
A: So the hill is fiery. (*Conclusion*)
B: Of course! (*Removal of doubt*).

The scheme is artificial at some points, as witness the vapidity of the remarks I have had to write in for the second speaker. However, there can now be little doubt that Gautama's scheme is aimed at representing the *presentation of an argument to others*; that is, at Rhetoric or Dialectic, not pure Logic.

A number of writers from the fifth and sixth centuries made a

[1] *Sophistical Refutations*, 165b 4. Roger Bacon produces a startlingly similar phrase in defining dialectical probability: 'The probable is that which seems the case to everyone, and about which neither the crowd nor the wise hold a contrary opinion': *Sumule dialectices*, p. 313.

distinction between *svārtha*, 'inference for oneself', and *parārtha*, 'inference for others', the former being possibly non-verbal. The distinction passed into the Indian logical tradition, which only much later, and then incompletely, developed a 'pure' Logic with a detachment from practical application, of the kind that has been dominant in the West.

The apparently repetitive character of the fourth and fifth members of Gautama's syllogism has been justified in another way. A tenth-century logician, Vācaspati Miśra, wrote[1]

> The *Conclusion* thus is not the same as the *Thesis:* the latter puts forward the fact only tentatively, as requiring confirmation by the reasoning with the aid of the *Reason* and the *Example*, while the former puts it forward as one fully established, and thus precluding the possibility of the truth being contrary to it. This cannot be done by the *Thesis*; as, if it did, then the rest of the members would be entirely futile.

Is a tentative thesis of the same 'form' as an established conclusion? Jhā sees their identification as the source of the paradox that every syllogism is question-begging, and the distinction between them as resolving it.

Turning now to fallacies, it will be enough to discuss, first, the treatment given by Gautama in the *Nyāya sūtra* and, second, the gradual change that took place in the tradition thereafter.

Gautama refers his fallacies to the 'reason', or second member of his syllogism. Vātsyāyana says that fallacious reasons 'are so called because they do not possess all the characteristics of the true reason, and yet they are sufficiently similar to the true reason to appear as such'. Gautama briefly and without detail or justification lists five classes of them. (1) First, a reason may be 'erratic' or 'inconclusive': Vātsyāyana's example is 'Sound is eternal because it is intangible', where in fact some intangible things are eternal and some not. (2) A reason may be 'contradictory' if it is in contradiction to something the proponent has already accepted or is known to hold. Vātsyāyana gives the example of two Yoga doctrines 'The world ceases from manifestation, because it is non-eternal', and 'The world continues to exist, because it cannot be utterly destroyed': these cannot both be right, and

---

[1] Quoted by Jhā, *Gautama's Nyāyasūtras*, p. 72, from the *Tātparya*.

once the first is accepted the second is fallacious, since its reason contradicts the earlier reason. (3) A reason may be 'neutralized' if, instead of leading to a decision about the thesis, it leaves matters undecided. It may do this because it is actually only a repetition of the thesis, and Vātsyāyana's example (p. 90) is 'Sound is non-eternal because we do not find in it the properties of the eternal thing'. (4) A reason may be 'unknown' or 'unproved'. 'Shadow is a substance, because it has motion' is of this character, says Vātsyāyana (p. 92), because it is not known whether a shadow has motion: 'Does the shadow move, like the man? or is it that as the object obstructing the light moves along, there is a continuity of the obstruction . . . ?' (5) Finally, a reason may be 'inopportune' or 'mistimed'. What Gautama means by this it is impossible to guess, but if commentators are to be believed it could be something to do with the fact that thesis and reason, as tensed statements, are true one at one time and one at another but not both together, as with Boethius's Fallacy of Different Time. Alternatively it could be just that the members of the syllogism are in the wrong order and, perhaps, that the thesis proves the reason rather than vice versa.

This list of five classes of fallacy formed the basis of various elaborated classifications of later writers, much in the way Aristotle's list did in Europe. If we suspend consideration of the impenetrable fifth one they are all, in a broad sense, 'formal': either the never-stated major premiss, the rule that is needed to justify the passage from reason to thesis, is false, or it is tautological, or the reason itself is unproved or elsewhere contradicted. It is true that some of the objections would not have been classified as 'formal' in the sense required in the West and which derives from Aristotle's *Prior Analytics*: the first class would be simply a case of false (suppressed) premiss, the second would be a case of a possibly-valid inference open to objection *ad hominem*, and the other two would be varieties of question-begging. None of them, however, involve variations in the meanings of words or phrases as in Aristotle's Fallacies Dependent on Language, and the kinds of dialectical irregularity that are involved are of a kind that can easily be provided with a formal analysis.

The *Nyāya sūtra*, however, also has a good deal to say about other kinds of logical fault or error. The section on fallacies had started out with a definition of three kinds of 'controversy'; namely, (1) Discussion, which is (p. 80) 'the putting forward (by two persons) of a conception and counter-conception, in which there is supporting and condemning by means of proofs and reasonings . . . carried on in full accordance with the method of reasoning through the five members'; (2) Disputation, which is discussion 'in which there is supporting and condemning by means of Casuistry, Futile Rejoinders and Clinchers'; and (3) Wrangling, which is disputation that is inconclusive. From the order of treatment it appears that the fallacies we have so far been discussing are such as occur primarily in the first of these three kinds of controversy and, though to be condemned, are of a lesser order of evil than what follows. Disputation and Wrangling, we are told, may be employed to keep up our zeal for truth 'just as fences of thorny boughs are used to safeguard the growth of seeds', and are of use against people who will themselves not argue properly.

Casuistry is of three kinds, of which the first is no more nor less than Equivocation, though the examples that Vātsyāyana gives of it make it equivocation of a particularly trivial kind: the word *nava* means alternatively 'new' or 'nine', and when someone says 'That boy has a new blanket' the casuist says 'No, not nine blankets, only one'. He goes on, however, to mention the circumstantial ambiguity of words, with examples like those of Sextus (p. 98):

> . . . when such expressions are used as – 'take the *goat* to the village,' 'bring *butter*', 'feed the *Brāhmaṇa*' – every one of these words ('goat', 'butter' and 'brāhmaṇa') is a general or common term and yet it is applied, in actual usage, to particular individuals composing what is denoted by that term; and to what particular individuals it is applied is determined by the force of circumstances; . . .

The casuist can put the wrong *denotation* on a word, and this is to be regarded as equivocation also.

The second kind of Casuistry is 'Generalising Casuistry', which consists in taking a speaker's words more generally than he intended them, and is slightly reminiscent of *Secundum Quid*. Someone who says 'Learning and character are quite natural to a *Brāhmaṇa*, does not necessarily intend 'delinquent' *Brāhmaṇas*,

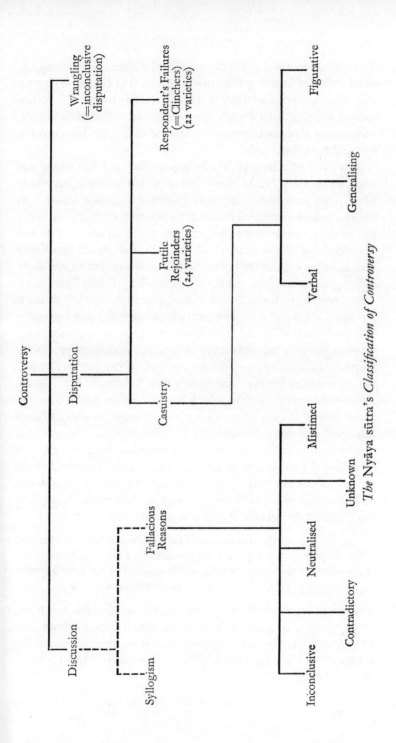

The Nyāya sūtra's Classification of Controversy

those who have not gone through all the rites and ceremonies, to be included and it will be casuistical to take him as doing so. The third kind of Casuistry is 'Figurative Casuistry' which consists in shifting the primary meaning of an opponent's words. Vātsyāyana discusses in some detail the difference between the first and the third.

What Jhā translates as 'Futile Rejoinders' and 'Clinchers' are dealt with in the *Nyāya sūtra*'s last, fifth book which has sometimes been regarded – on what evidence I do not know – as written independently of the rest. This part of the book has had a bad press, for reasons we shall explore in a moment; but its aim has clearly been misunderstood, and we shall find a syndrome that is by now familiar from our study of the fate of the early Aristotle. The list (pp. 502 ff.) of twenty-four 'Futile Rejoinders' – the translation seems a little exotic, since the Sanskrit word is just *jāti* – is a list of ways in which a piece of syllogistic reasoning may be 'equalized': that is, of ways in which an opponent may bring arguments that balance or neutralize the original reasoning without challenging it on its own ground, in the way the author would consider proper. The twenty-two 'Clinchers' (pp. 540 ff.) – the word is *nigrahasthāna* which has sometimes been translated 'Respondent's Failures' are ways in which the *proponent* of a thesis can spoil his case with dialectical shortcomings. The two lists can be seen as complementary if we imagine them as a manual of debating tactics, and as setting out possible dialectical faults of the opponent and proponent, respectively, in a debate. Stcherbatsky, in his well-known book on Buddhist Logic, does not mention the first list at all but says of the second[1]

> The Manual on the Respondent's Failures was evidently a manual for the judge, its composition the result of a long experience in the practice of the art of debating, which resulted in the establishment of a system of type-instances and laws regulating the debate.

The term 'respondent', borrowed from the Western tradition, presupposes a question-and-answer form of debate which may or may not have been usual; and there is no direct reference in Gautama or Vātsyāyana to a 'judge'. Nevertheless this properly represents the tone of the fifth book.

[1] Stcherbatsky, *Buddhist Logic*, p. 340.

The 'Futile Rejoinders', necessarily too briefly described to do them full justice, and with the omission of a few which are obscure or repetitious, are as follows:

[1 and 2] argument for a contrary of the thesis (i.e. without destroying the original one),
[3] arguing that the example proves other things as well,
[4] arguing that the subject of the thesis lacks some properties of the example,
[5 and 6] arguing that the property in question is uncertain in the example, or less certain than in the subject,
[7] arguing that the example is contingent,
[8] presenting the example as equally well to be proved from the thesis (i.e. the claim that the example begs the question),
[9 and 10] arguing that the reason is 'united with' the thesis, and cannot prove it; or that it is unconnected with it, and cannot prove it,
[11] arguing that the reason is itself in need of proof,
[14] independently throwing doubt on the thesis,
[17] arguing by *reductio ad absurdum*, on the presumption of consequences the proponent might not grant,
[18] objecting generally to the method of argument by example,
[20] arguing that the thesis is in fact known to be false,
[21] (perhaps) arguing that the thesis is not known not to be false.

It will be seen that even arguments quite good in themselves, as in [1] and [2] and, above all, [20]!, are regarded as 'futile' when they come *in answer to* the original reasoning. This does not mean, presumably, that the opponent should not put them forward but only that, in neutralizing the original reasoning, they are themselves neutralized as well. It is not the opponent's job to establish a counter-thesis at all.

'Clinchers' or 'Respondent's Failures' are easier to summarize: the following list partly follows Stcherbatsky. The proponent of an argument *may* be criticized for:

[1] annihilating his own thesis by an unsuitable example,

[2] shifting to another thesis,
[3] giving a reason that contradicts the thesis,
[4] abandoning the thesis,
[5] changing the reason originally given,
[6] irrelevancy,
[7] giving meaningless sounds as a reason,
[8] giving an assertion that is unintelligible even though stated three times,
[9] giving a syllogism such that there is no connection between the members,
[10] stating the members in the wrong order,
[11, 12] reasoning that is incomplete, or redundant,
[13] repetition,
[14, 15] failure to restate, or understand, opponent's objection,
[16] admission of ignorance,
[17] breaking off the debate (thus conceding defeat),
[18] admission of a flaw in his reasoning,
[19, 20] neglecting to rebuke the opponent when necessary, or doing so when not necessary,
[21] irregular discussion.
[22] fallacious logical reasons.

Only a few comments need be made on these lists, which reveal a preoccupation with orderly debate as strong as anything in Aristotle. The first concerns item [22] of the second list, whose inclusion is quite out of place with the classification-scheme suggested earlier by Gautama and is perhaps evidence of separate origin, though it could as easily be due to a conflict between two conceptions in the mind of a single writer. The second is the startling similarity between some of the individual items and some of Aristotle's. Two hypotheses are possible: that there was contact between, or a common origin of, the two traditions; or that formal debate is an important or necessary ingredient of any intellectual culture at a certain stage of its development, and is the driving force behind the development of Logic. They are not, of course, mutually exclusive.

The precise shape that formal debate took at the time of the *Nyāya sūtra* can be only dimly guessed at. There do exist records of formal debates from Buddhist sources at an earlier period, the

THE INDIAN TRADITION

reign of King Aśoka, about 255 B.C.[1] They seem to consist of phases in which the two disputants alternately take the floor, and each phase consists of the statement of a single argument, perhaps preceded by a number of clarificatory questions addressed to the opponent. I quote (with my own minor amendments) the first two phases of a debate between monks of rival views concerning the reality of the soul:[2]

PRESENTATION

SCEPTIC Is the soul known in the sense of a real thing?
SUBSTANTIALIST Yes.
SCEPTIC Is the soul known in the way a real thing is known?
SUBSTANTIALIST No, that cannot be said.
SCEPTIC Acknowledge your defeat:

(i) If the soul is known in the sense of a real thing, then, good sir, you should also say that the soul is known in the way a real thing is known.

(ii) What you say is wrong, namely (*a*) that the soul is known in the sense of a real thing, but not (*b*) known in the way any other real thing is known.

(iii) if (*b*) is not admitted (*a*) cannot be admitted either.

(iv) In admitting (*a*) but denying (*b*) you are wrong.

REJOINDER

SUBSTANTIALIST Is the soul not known in the sense of a real thing?
SCEPTIC No, it is not.
SUBSTANTIALIST Is it unknown in the way a real thing is known?
SCEPTIC No, that cannot be said.
SUBSTANTIALIST Acknowledge the rejoinder:

(i) If the soul is not known in the sense of a real thing, then, good sir, you should also say that the soul is unknown in the way a real thing is known.

(ii) What you say is wrong, namely (*a*) that the soul is not known in the sense of a real thing, but not (*b*) unknown in the way a real thing is known.

(iii) If (*b*) is denied (*a*) cannot be admitted either.

(iv) In admitting (*a*) but denying (*b*) you are wrong.

A judge, elected by the assembly (of monks), presides. Whether

---

[1] See Vidyābhūṣaṇa, p. 234.
[2] Vidyāhbūṣaṇa, pp. 235–6, quotes from the *Kathāvatthu*; see also Bocheński, p. 421.

this is the kind of debate Gautama had in mind remains, however, conjectural.

Most of the works written directly in the Nyāya tradition are actual commentaries on the *Nyāya sūtra*, or commentaries on commentaries. Some of them ring changes on the doctrine without fundamentally altering it. Many invent, like Aristotle's commentators in medieval times, elaborate fine subdivisions of the various categories and, at the hands of Uddyotakara (seventh century), the original fivefold division of fallacies reaches the record grand total of 2,032 subdivisions. We need not investigate these subtleties. Many of the histories subdivide the tradition by religious groupings, so that we are presented in the sixth century with the separate 'Logics' of Praśastapāda (Hindu, Vaiśeṣika tradition), Siddhasena (Jain), and Diṅnāga (Buddhist), but the divisions refer more to epistemological doctrines than to the purely logical topics that here interest us.

Diṅnāga is the most original of these writers and must here do duty for the others. Although he is not uninterested in controversy his theory of inference displays a move away from Dialectic: his 'syllogism' drops the last two members, leaving only *Thesis*, *Reason*, and *Example*, though he adds the concept of a *Counter-example* in an attempt to eke out the deficiencies of this argument by analogy: 'The hill is fiery, because it is smoky, like a kitchen, unlike a lake'. It is not clear whether the Counter-example is meant to support the converse of the general argument or merely the contrapositive. He explicitly rejects Testimony as a basis of argument. On fallacies he gives a fourteen-fold classification of defects that is completely formal in character and tied to his theory of the syllogism: fallacies may infect the thesis, the reason or the example, and these in various ways. The reason, for example, may be 'too wide' or 'too narrow', the reason or example may themselves be uncertain, and so on. Diṅnāga does not attempt to classify controversy as the *Nyāya sūtra* does, does not mention Casuistry, gives a list of only some of the Futile Rejoinders, and ignores Respondent's Failures.

Not only in Diṅnāga, but even in such of his contemporaries and successors as remained close to the Nyāya teachings, the second or 'contradictory' category of fallacy ceased to represent the circumstance of an opponent's contradicting his own doctrines and came to represent the Fallacy of giving a reason that tends to contradict, rather than support, the proffered thesis. Hence, this too became formal in character. The obscure fifth category of fallacy was variously interpreted but tended to become a rag-bag of cases in which the truth of one or another member of the syllogism was regarded as questionable for extraneous reasons.

The move towards a formal, deductive logic continued after Diṅnāga, and Uddyotakara gave a version of the syllogism in which the Example was at last replaced by the statement of a general rule. The final turning-point came in the fourteenth century with the work of Gaṅgeśa and the growth of what has become known as the *Navya Nyāya*, or New *Nyāya*, school. Gaṅgeśa mentions Casuistry, Futile Rejoinders, and Respondent's Failures only to criticize them, and this seems to be the last point in the history of Indian Logic at which they have seriously been considered at all.

The development of Logic out of a theory of debate, and its ultimate repudiation of its origins, thus seems ultimately to have taken the same course in India as in the West. That the classical Indian list of fallacies, like that of the West, is still studied despite the loss of its principal rationale is support for the view that the study of fallacies in both places has a function different from its ostensible one.

CHAPTER 6

Formal Fallacies

The most remarkable feature of the history of the study of fallacies is its continuity. Despite the waves of disinterest and rebellion that phase and punctuate it, and despite fundamental changes in underlying logical doctrine, the tradition has been unquenchable. The lesson of this must be that there is something of importance in it. It is true that most of what appears in the modern books has very little relation to what Aristotle wrote, and it may also be true that much of it is incoherent; but the writers of the modern books are, as the saying goes, all honourable men. In short, we have some explaining to do. If we now turn, as is appropriate, from a historical account to an analytical logical one, it must be in the attempt to answer the questions: what, in the tradition, is worth keeping, and how can it be separated from the parts that should be thrown away?

If Aristotle could be taken as our norm we could concentrate on producing a new and error-free *Sophistical Refutations*, shorn of irrelevant outgrowths. This would, in some ways, be better than nothing; but we would miss out on all the lessons that the misconceptions of History are themselves capable of teaching us. Distorted versions of Aristotle's doctrine have continued to be produced, we must assume, only because they have been needed. To understand the tradition thoroughly we should see what is common and what is variable, in all the treatments of all the ages. If any age is to be accorded favouritism it should be our own.

On the other hand, there does not seem to be much point in producing *any* new theory of fallacies in the same vein as the old.

There have been too many already. Least of all should one aim at producing a new *classification* of the old material, yet another tree. Classification is sometimes a useful preliminary in dealing with unstudied material, and sometimes – not always – it is the upshot of a successful investigation. Our concern, however, is with the middle part of the study, and so many have failed with this that it should be clear by now that they have been asking the wrong questions. One of the main reproaches that could be brought against the study of fallacies is that it has always remained an appendage, insecurely connected to the main part of Logic. A new classification of fallacies does nothing to remedy this; and, if the subject cannot be brought into closer relation with the rest of Logic, a radical reappraisal, either of the study of fallacies, or of the rest of Logic, is called for.

To consider a possible parallel with our situation, let us suppose that Aristotle had never invented his theory of syllogisms and that, in place of a formal theory of inference, we were left with the loose list of hints for conducting and winning arguments that have come down in the tradition of his *Topics*. Since arguing is sometimes a serious business, and since any theory of inference is better than none at all, it would be right and proper under these circumstances that Topics should continue to be written up in every Logic book and discussed with every student; but since, as with Fallacies, there is no clear pattern in any list, everyone could have his own system of classification and his own variant descriptions. The theory of inference, as we in fact know, can be presented more systematically and satisfactorily than this, and Topics, though carried for many centuries as an appendage, have finally disappeared from logic books. As a theory of inference, they were important only so long as there was nothing better. We shall similarly superannuate fallacies when we see what their function is and how it can be fulfilled in a modern idiom. History, of course, cannot justify us in being sanguine about breaking through; but that is no excuse for not trying.

Our first question concerns how much can be accomplished within Logic's existing framework, modern Formal Logic: we owe it to fellow logicians to do what we can to answer this. It should immediately be added that there is much in Logic that has not changed since Aristotle. Notably, (1) Logic is conceived as having rules expressible in *schemata*, involving variables, whose logical properties are independent of what is substituted; (2) Logic produces truths, or rules, which are common to all other disciplines, and hence of a different order from (higher or lower than) those of other disciplines; (3) the logical unit is the *proposition*, and its leading logical property is its *truth-value* (truth or falsity), whence it is associated with the concepts of negation, of contradiction and 'excluded middle'; (4) there is a primary concern with rules of deduction (or inference, or implication), conceived as proceeding *from* one or more premisses *to* a conclusion; essentially reflexive, non-symmetrical, and transitive; (5) *proof* is conceived as a kind of deduction, knock-down, non-cumulative; (6) speaking generally, the theory is exclusive in the sense that reasoning processes of other kinds – inductive, extrinsic, emotive – are accorded lower status; (7) the theory is impersonal and context-free. This is an impressively long list of seldom-questioned presuppositions; and it should be clear by now that many of them will need to be challenged in the process of making sense of the fallacy-material and, perhaps, will need to be sloughed off one by one or in combinations. For the present chapter, however, let us see what we can do by keeping them intact.

For, although there are reasons for thinking that some of the phenomena of Logic are outside the bounds of the kind of formal system in which logicians, ancient or modern, have preferred to formulate their theories, there are many who do not think this and whom we must meet on their own ground. Take, for example, the Fallacy of Begging the Question: it can well be imagined that a logician might approach an analysis of this Fallacy by inventing a new formal concept of implication such that question-begging inferences were exposed by it. In the resulting system it would be necessary, among other things, to reject – that is, to deny the validity of – the *schemata* 'p implies p' and 'p and q together imply q'. It is not clear that there is anything impossible about this, and we need to discuss whether it would count as a solution or part-solution to the problem of giving an acceptable modern account of the Fallacy. Again, no one, so far as I have been able to find out, has ever tried to produce a formal system in which arguments involving equivocation can be represented. To do so would be a little odd since one of the usual aims of formal systems is to be unambiguous; but, again, it is not clear that we cannot learn anything from a system containing systematic ambiguities.

The task of the present chapter is twofold. In the first place I ask seriously whether any general and synoptic theory of fallacy can be extracted from formal studies or stated in formal terms. Since the answer to this appears to be in the negative, I shall secondly consider whether there are formal analyses of particular Fallacies in the traditional list; either (*a*) in more-or-less orthodox formal terms, or (*b*) within formal theories that can be constructed specially for the purpose. The traditional fallacy-material will determine broadly the scope and meaning of the word 'fallacy'.

Our first question, then, is whether any general and synoptic formal theory of fallacy is possible. We shall need to be a little clearer about what the word 'formal' means. In particular, it is not at all clear what a 'formal fallacy' is. This is partly because it is not clear whether the rules of Formal Logic are supposed actually to declare certain arguments invalid, or merely to declare certain ones as valid and leave the rest open; but the trouble runs deeper and is concerned with the relationships of formal languages or canonical forms to the natural languages in which Logic must be put to practical use. Within a *formal language* it is generally clear enough which arguments are formally valid; but an ordinary-language argument cannot be declared 'formally valid' or 'formally fallacious' until the language within which it is expressed is brought into relation with that of some logical system.

In retrospect, it looks as if the concept had more relevance a hundred years ago; so long, that is, as Aristotle's theory of syllogisms was the accepted theory of inference and Logic and educated language had grown together to the point where problems of interpretation were at a minimum. Within this developed tradition it should have been possible to achieve a limited agreement about the distinction between formal fallacies and others. Yet history does not give us evidence of such agreement. Certainly, Aristotle never achieved a clear grasp of any such distinction. The earliest account of formal fallacies is in the writings of Cassiodorus, contemporary of Boethius in the sixth century, whose short chapter on 'paralogisms'[1] is a survey of the ways in which syllogisms may contravene Aristotle's rules; but this is not accompanied by any survey of fallacies of other kinds and gives us no criterion of differentiation. The medieval treatises generally keep their discussions of the syllogism and their discussions of sophisms rigidly separate; and, although the examples of the latter are generally in syllogistic form ('Everything that runs has feet; the river runs: therefore, the river has feet') they are almost always (like this example) valid by formal criteria and fallacious only because additional, evidently non-formal criteria need to be brought in. The exceptions are the examples of Consequent and Non-Cause. That there is a distinction to be drawn between arguments which contravene the rules of the syllogism and those which, although they do not do so, are invalid for other reasons is obvious enough. However, a fallacy, as we must remind ourselves, is an argument which *seems* valid but is not; and many invalid arguments have no appearance whatever of validity. The 'formal' validity of a fallacy-example may be what provides it with an innocent face.

The distinction between the 'form' and 'content' or 'matter' of a proposition or argument is straightforward when, as in Aristotle, schematic letters are used for some of the terms. Aristotelian syllogisms are written in textbooks in forms such as 'If all Bs are Cs and all As are Bs, then all As are Cs'; and this is the pure form or *schema* of a syllogism, which can be given content only when some substitution is made for the letters – say 'Greeks', 'men', and 'mortals' for 'As', 'Bs', and 'Cs' respectively. The medieval writers were merely codifying this distinction when they distinguished the content-words or *categoremata* from form-words or *syncategoremata*; the latter being words like 'all', 'some', 'each', 'no', 'not', 'and', 'or', 'is', 'only', and 'except', whose function in a sentence is not referential but structural. Modern logicians tend still to make a similar distinction: Russell divided his symbols into 'logical variables' and 'logical constants', and it is only in certain rather sophisticated special con-

[1] Migne, *Patrologia Latina*, vol. 70; cols. 1194–6.

texts that the distinction becomes clouded.[1] Formal Logic, then, is the Logic that can be studied in terms of form, exclusively of content, and has a long history.

The idea of distinguishing certain types of argument as 'formal fallacies' is, however, a relatively recent one dating from Whately, who took up a hint in Aldrich.[2] We should have a look at what Whately says since, besides having a lesson for us as to the place of the concept of 'formal fallacy', it contains an interesting attempt to provide a unified theory of formal fallacy and of Equivocation.

A *syllogism* is a three-term three-proposition argument such as 'All men are mortal; all Greeks are men: therefore, all Greeks are mortal', or 'All lawyers are graduates; some graduates are dishonest; therefore, some lawyers are dishonest'. We shall not be concerned here with syllogisms containing modal terms which, although treated by Aristotle, have generally been ignored. The second of the two examples given is invalid, since it is in theory possible that the dishonest graduates referred to should all be from disciplines other than Law and that all products of the latter be paragons of virtue. The *terms* of a syllogism are the various subjects and predicates, e.g. 'Greeks', 'men', and 'mortal': the *middle term* is the one that appears in both premisses and is the means, as it were, by which the conclusion is drawn, e.g. in this example 'men'. The *major term* is the one that appears as predicate of the conclusion, the *minor term* the one that appears as subject. The *major premiss* is the premiss containing the major term.

Aristotle's theory of inference divides syllogisms into various *figures*, or patterns of occurrence of the terms, and *moods*, or patterns of occurrence of the words 'All', 'No', 'Some', and 'Not all'; and then reduces all moods to a few in the first figure, which were taken to be 'perfect'; that is, not in need of explanation. Generations of students, however, learned the valid moods by rote. A simple set of rules of validity was finally produced in the later Middle Ages, based on the concept of Distribution.

In theory a term is said to be *distributed*, at a given occurrence

---

[1] I have in mind the 'protothetic' of Leśniewski, which introduces variable logical operators: see, for example, Prior, *Formal Logic*, p. 66.
[2] Whately, *Elements of Logic*, Book III. See the passage quoted from Aldrich above, p. 49.

in a given proposition, when it refers in its context to *all* the members of the class of objects it is capable of denoting. In 'All men are mortal' the term 'men' is distributed, the term 'mortal' not; the latter because the proposition is not necessarily about all mortals. Generally in a proposition of the form 'All *A*s are *B*s', the subject-term is distributed. In the case of 'No *A*s are *B*s', both terms are distributed, since it is implied of every *A* that it is not a *B* and of every *B* that it is not an *A*. In the case of 'Some *A*s are *B*s' and 'Not all *A*s are *B*s' there is no good case for saying of any of the terms that they are distributed, but the theory requires that we say that the term '*B*s' is distributed in the latter. Since the statement that, for example, not all radioactive elements are metals does not enable us to deduce anything about all metals, or about any given metal, the common explanation of Distribution needs to be modified. The reader may be referred elsewhere, however, for details.[1]

Rules of validity for syllogisms stated in terms of Distribution have been a regular feature of Logic textbooks since the seventeenth century. There is, for example, a list of twelve rules, summed up in a Latin verse, in Aldrich.[2] Sets as large as this are always highly redundant, in the sense that some rules may be deduced from others. It is convenient to pick up the story with Whately, who gives a set of six (from *Elements of Logic*, Bk. II, ch. III, §2):

1. Every syllogism has three, and only three, terms.
2. Every syllogism has three, and only three, propositions.
3. The middle term must be distributed at least once.

---

[1] For a recent eruption of controversy over Distribution see Geach, *Reference and Generality*, and review by Quine in *Philosophical Review*, 73 (1964), pp. 100–4. Some sense can alternatively be made of Distribution within a Logic of quantified predicate terms as in Hamilton, *Lectures in Logic*, vol. II, pp. 257–323. Hamilton reviews the historical precedents: cf. William of Sherwood, *Introduction to Logic*, pp. 38–9.

[2] The verse is (p. 75):
> *Distribuas medium; nec quartus terminus adsit.*
> *Utraque nec praemissa negans, nec particularis.*
> *Sectetur partem conclusio deteriorem.*
> *Et non distribuat, nisi cum praemissa, negetve.*

Mansel gives an earlier verse that appeared in some later editions of Peter of Spain's *Summulae*. See also Kneale, pp. 272–3.

FORMAL FALLACIES

4. No term must be distributed in the conclusion which was not distributed in one of the premisses.
5. From [two] negative premisses you can infer nothing.
6. If one premiss is negative, the conclusion must be negative.

Besides providing a basis for determining whether a given syllogism is valid, and hence a definition of 'formal fallacy', these rules give us a system of classification of formal fallacies. If rule 3 is broken the Fallacy of the Undistributed Middle Term is committed; and if rule 4 is broken, the Fallacy of Illicit Process of the Major Term, or of the Minor Term, depending on which term is involved. The Fallacy of Four Terms infringes rule 1. There are no accepted names for breaches of the other rules, but it would be easy enough to invent some.

Some twentieth-century books[1] omit the first two rules on the grounds that they represent part of the definition of a syllogism, and must hold, as it were, even for invalid syllogisms: they are not, like the other rules, part of the *differentia* of valid syllogisms from invalid. This is reasonable; but we must first consider an attempt by Aldrich and Whately to conflate the Fallacy of Four Terms with that of Undistributed Middle. In fact the tag 'Fallacy of Four Terms' has usually been applied, as described above,[2] to arguments involving an *ambiguous* middle term. Whately gives (Bk. II, ch. III, §2):

'*Light* is contrary to darkness;
Feathers are *light*; therefore
Feathers are contrary to darkness.'

If we read the three terms as 'light things', 'things with a property contrary to darkness' and 'feathers', this is of the form

All B_1s are Cs
All As are B_2s
Therefore, all As are Cs.

where we have written 'B_1' and 'B_2' for the two different occurrences of the term 'light things', on the ground that it has different meanings. Are B_1 and B_2 really one term or two? Dropping the notion of a 'term' it seems sensible to say simply that 'light' is *one* word with *two* meanings, and that the fallacy is not one of Four Terms but of Equivocation.

[1] e.g. Cohen and Nagel; see p. 79.   [2] pp. 44–5.

Whately distinguishes clearly enough between Four Terms and Equivocation in his tree of Fallacies given above in chapter 4, but he is prepared to go along with Aldrich in presenting an alternative account in which not merely these cases but also cases like

> Some animals are beasts
> Some animals are birds
> Therefore, some birds are beasts.

are cases of Four Terms. The middle term 'animals' is undistributed in both premisses and it would be quite easy to rule the inference invalid under Whately's rule 3. Aldrich, however, besides having rules like all of Whately's, has also:

> If the middle term is ambiguous, nothing follows.

and

> An undistributed middle term is ambiguous.

In respect of the second of these he says (pp. 71–2):

> Thus let B be a common term divisible into b and β. It follows that b and β are opposed; and yet we may truly say both 'Some B is b' and 'Some B is β'. Therefore, 'Some B' is an ambiguous middle term.

This gives him even more resources: he can rule the above-mentioned syllogism invalid not only in the same two ways as Whately, but also using the reasoning associated with the two rules given. What is interesting is not so much the fact that he can invoke more than one rule as the fact that Undistributed Middle is *identified* with Four Terms.

It would be possible to extend a similar analysis to breaches of Whately's rule 4, and even, with some special gyrations, to breaches of rules 5 and 6. In the case of most breaches of rule 4 it is obvious that a conclusion about all members of a class is being drawn on the basis of premisses about only some of them, and the conclusion will hence say something about certain things that are not mentioned in the premisses. Hence we can regard the appropriate term – major or minor, whichever it is – as ambiguous. Where negative premisses or conclusion occur we can replace them by affirmative ones if we replace certain terms by their complements; but will sometimes find that a term and its complement will need to be reckoned as two. Thus, if we allow Aldrich's analysis, all invalid syllogisms can be regarded as cases of the Fallacy of Four Terms. Unfortunately we must not allow it.

When Aldrich says that 'Some B' is an ambiguous middle term he departs from the concept of a *term* by including the word 'Some': 'B' represents a term on its own, whatever is prefixed to it. Moreover, the term 'B' represents can be ambiguous on its own account; as, for example, 'light things' is ambiguous. Aldrich has two sorts of ambiguity, namely (i) that which can occur in certain terms in particular, and can occur in these terms irrespective of their position in a proposition, and (ii) that which is a function of the position of a term in a proposition and is independent of the term itself. The two sorts are quite independent of one another ... and only the first would normally be called 'ambiguity': the term 'animals' in 'Some animals are birds' is ambiguous only in Aldrich's idiosyncratic second sense, and not in the sense in which we ordinarily use the word. Consequently it is desirable to reserve the word 'ambiguous' for the first sense, and call terms of the second kind simply 'undistributed'.

So let us return to Whately's set of six rules, and strike out the first two: they should be replaced by a definition of 'syllogism' that applies merely to the form of the argument, independently of whether it is valid or invalid. Rules 5 and 6 should now be looked at; for these two rules do not operate independently in the way the others do, but rather partition the field between them. They can be rewritten together in the form:

5 '. There is an affirmative conclusion, a negative conclusion or no conclusion at all according as both premisses are affirmative, or only one, or neither.

The rules 3, 4, and 5' provide us with a satisfactory modern theory of validity for syllogisms.[1] The Fallacy of Undistributed

[1] Strictly, only for 'classical' syllogisms; that is, excluding empty terms. If 'empty' terms are permitted, some of these must be counted invalid and an additional rule is required: 'If the conclusion is particular (i.e. existential), a premiss must be particular.'

Middle occurs if 3 is broken, the Fallacy of Illicit Process if 4 is; and for breaches of 5' we may coin the name 'Fallacy of Negativity', perhaps permitting sub-divisions. The syllogism

> All soccer fans are excitable.
> Some tympanists are excitable.
> Therefore, some tympanists are soccer fans.

commits the Fallacy of the Undistributed Middle; the syllogism

> All tropical countries are overpopulated.
> No tropical countries are industrialised.
> Therefore, no industrialized countries are overpopulated.

commits the Fallacy of Illicit Process (of the Major Term, 'overpopulated'); and the syllogism

> No inert gases form chemical compounds.
> Not all inert gases are found in the atmosphere.
> Therefore, not all things found in the atmosphere form chemical compounds.

commits one form of the Fallacy of Negativity. In the last example we must be careful, of course, not to read an affirmative proposition, 'Some inert gases *are* found in the atmosphere', into the second premiss.

These are all formal fallacies, and they are all uniquely classifiable. However, a syllogism can break more than one rule at once; and, with a little ingenuity, we can succeed in breaking all three. The syllogisms

> Some doctors are dentists
> Some dentists are diplomats
> Therefore, no diplomats are doctors.

and

> Not all manuscripts are irreplaceable
> Some manuscripts are indecipherable
> Therefore, all indecipherable things are irreplaceable.

both, in different ways, pull off this feat. The first has undistributed middle term, illicit process of both major and minor, and a negative conclusion with no negative premiss; the second has undistributed middle, illicit process of the minor term 'indecipherable', and an affirmative conclusion with one premiss negative. What this means is that, although the set of three rules is quite adequate to define validity and hence formal fallacy, it does not give us a *classification* of fallacies, in the sense of a division into mutually exclusive categories; unless we are content to count each possible combination of ways the rules may be broken as generating a different category, in which case there would be seven categories altogether.

Generally, if we want a set of rules to act both as a set of jointly sufficient and individually necessary rules of validity and as a classificatory system giving mutually exclusive categories for fallacies, we shall need to impose some additional constraints on the rules. We shall want them, in particular, to be so formulated that it is impossible to break more than one rule at once. If we do not do this there will be no special sense in which the breaking of a rule gives rise to a particular kind or variety of fallacy; and, perhaps more important, no special way in which the study of formal fallacies contributes to the theory of validity: the *classification* of formal fallacies will be pointless. We should, then, explore this requirement of additional constraints.

Two propositions so related that they cannot both be false are said to be *subcontrary* to one another; and this concept is easily extended to rules as well as propositions. Consequently we can express our new requirement on sets of rules as follows: The various rules which go to make up a set should, if we want a one–one relationship between rules and fallacy-types, be mutually subcontrary. It should, of course, be possible to satisfy them all at once; but it should never be possible to break more than one at a time.

It so happens that it is possible to satisfy this requirement almost trivially. Let P, Q, and R be propositions representing the states of affairs stipulated by three rules which together form a set of the kind we have been considering: that is, let the three rules be 'P must be the case', 'Q must be the case', and 'R must be the case'. We shall suppose initially that our additional stipulation is not satisfied, and that P and Q are not subcontraries and can be false together. Leaving P unaltered we shall transform Q in such a way that it can still be true at the same time as P, and so that P and Q *together* still specify the same thing, but so that Q cannot be false if P is. Briefly, we replace Q by Q' which says

Either P is false or Q is true (or both), so that our first two rules are now

> [1] 'P must be the case'
> [2] 'It must be the case either that P is false, or that Q is true (or both).'

The second of these appears to be capable of being satisfied by having P false, but, of course, we may disregard this possibility since the first could not be satisfied simultaneously. However, [2] can *fail* to be satisfied only if P is true and Q false, and hence not at the same time as [1].

If R is not mutually subcontrary to the others we may replace it by the proposition R':
Either P is false or Q is false or R is true.
so that the new third rule reads

> [3] 'It must be the case either that P is false, or that Q is false, or that R is true.'

This can fail only if P and Q (and hence Q') are true, and hence only if [1] and [2] are satisfied.

Consequently, we can obtain a set of three mutually-subcontrary rules for validity of syllogisms as follows:

> [1] The middle term must be distributed at least once.
> [2] Either the middle term must be undistributed; or any term distributed in the conclusion must be distributed in the premisses.
> [3] Either the middle term must be undistributed; or there must be a term distributed in the conclusion and not in the premisses; or there must be an affirmative conclusion, a negative conclusion or no conclusion at all according as both premisses are affirmative, or only one, or neither.

What these gain in discrimination they lose in perspicuity. We can hardly now dub breaches of the third rule just 'Fallacies of Negativity': they will have to be 'Fallacies of Negativity with Distributed Middle and Licit Process', and breaches of [2] will have to be 'Fallacies of Illicit Process with Distributed Middle'. There is something arbitrary and *ad hoc* about all this, and we realize that an important requirement of our problem – that its solution should be, in some sense, 'natural' – has been overlooked. If we cannot satisfy this requirement we should, perhaps, give up trying to produce a *classification* of formal fallacies altogether.

This conclusion is reinforced by a consideration of what would be involved in producing, for *modern* Logic, a set of rules of inference along the lines of the Distribution rules. Within the propositional logic that has grown up since Boole, validity of a formula is usually determined either by means of a deduction from axioms, or by means of an argument from truth-values. It would, perhaps, be possible to invent a set of rules, separately necessary and jointly sufficient, that would embody the essential features of either of these methods; but to do so would be artificial in the extreme and, in some sense, at variance with the spirit of the system. We should notice, in this connection, first, that sets of rules of 'Consequences' were produced in profusion in the later Middle Ages,[1] though without achieving either mutual independence of the rules or exhaustivity; and, secondly, that some nineteenth- and twentieth-century logicians[2] give us a small selection of these rules and use them to proscribe the Fallacies of 'affirming the consequent' and 'denying the antecedent'. Such treatments are not only too fragmentary, but also strangely ill-judged when set against the theory of validity of a Frege, a Russell, or a Quine. They would be even more out of place in dealing with inferences in the predicate calculus or set theory.

Let us clutch at what classificatory straws there are. Before turning to modern formal analyses of particular Fallacies, we should notice just one further piece of ancient doctrine which might be pressed into service. This is the doctrine of the seven logical relations between pairs of propositions. In its essentials the doctrine is from Aristotle (*Interpretation* 17b 16), who

---

[1] See Kneale, pp. 274–97; Bocheński, pp. 189–209. The text of an exceptionally large set is given by Ivan Boh in 'Paul of Pergula on Suppositions and Consequences', *Franciscan Studies*, 25 (1965), pp. 55–67.

[2] See, for example, Stebbing, *Modern Introduction to Logic*, pp. 103–8.

made the vital distinction between *contradiction* and *contrariety*: two propositions A and B are mutual *contradictories* if each is a precise denial of the other, but *contraries* if, although they cannot both be true, it is possible that they should both be false. The other five relations – *equivalence, superalternation, subalternation, subcontrariety*, and *indifference* – complete the systematic picture.[1] The names *superimplication* and *subimplication* are sometimes used in place of the second and third of these. It is easy to show that the enumeration is, in a suitable sense, exhaustive. The doctrine is equally applicable to statements in (reasonably normal) modern logical calculi, at least if we restrict ourselves to 'contingent' statements – that is, statements that are not themselves theorems or the negations of theorems – or if we find some other way of avoiding paradoxes of implication.[2]

Now, of the seven possible relations that A may have to B, two, superalternation and equivalence, are such that B may be formally deduced or inferred from A and, in these cases, the inference of B from A is formally valid. Two others, contradiction and contrariety, are such that we can deduce from A that B is false. Here, the inference of B from A would be *contravalid*. In the other three cases A does not imply B, but does not counter-imply it either. It is presumably of some importance to recognize that some formally invalid inferences are actually contravalid, and some not; and that formal rules themselves give us at least these two kinds of invalidity. The finer subdivision into two kinds of validity and, in all, five kinds of invalidity would

---

[1] They first appear in Apuleius: see Bocheński, *History of Formal Logic*, p. 140. A similar set of distinctions is made in Indian tradition by Diṅnāga (sixth century?), in his 'wheel of reasons': see Stcherbatsky, *Buddhist Logic*, vol. 1, pp. 320–7. Diṅnāga's ninefold division is actually isomorphic to a ninefold division that appears in Cohen and Nagel, *Introduction to Logic and Scientific Method*, pp. 55–6, and which includes cases of tautological and self-contradictory consequent (in Diṅnāga's case, universal and empty Reason): compare the fifteenfold table in my *Elementary Formal Logic*, p. 182. Diṅnāga, of course, knew nothing of Apuleius or Western tradition but makes several of our points for us.

[2] For example, it is applicable as between contingent statements in any system which contains the ordinary statement-calculus, and is consistent; or in any truth-functional system with a 'normal' implication and negation. We shall return later to some of the problems of implications between theorems and antitheorems.

not often be important but this, too, is implicit. In cases in which there is more than one premiss to an argument we can regard the premisses as conjoined into a single one.

Yet there is not a single one of the traditional headings of fallacy that is illuminated by this classification. Least of all does it have any relevance to Equivocation, Amphiboly, and the other Fallacies Dependent on Language; or to the *ad hominem* or *ad verecundiam*. Of the Aristotelian Fallacies Outside Language it certainly has no relevance to Accident, *Secundum Quid*, or Many Questions, and the detailed analysis of Non-Cause as Cause is too complicated to be summed so simply. It is just possible that a case could be made for saying that Aristotle had the relation of indifference dimly in mind when referring to Misconception of Refutation; and/or the relation of equivalence in the case of Begging the Question; and/or the relation of subimplication in the case of Consequent. At face value, however, none of his examples of any of them bear out these identifications directly, and all the Fallacies seem to have extra elements in them. We shall deal with them in more detail in a moment; but it is very clear that even if all the identifications were sustained we could not say that the theory of logical relations has more than minor relevance to the traditional list of Fallacies.

In fact, the answer to our first main question of this chapter – whether it is possible to give a general or synoptic account of the traditional fallacy-material in formal terms – seems to be a simple 'No'. This is not, by now, surprising. Fallacies have continued to occupy a place in textbooks largely because they introduce important considerations outside Formal Logic and supplementary to it. We might remind ourselves that some modern books[1] even refer to a large class of them specifically as 'Informal Fallacies'. The contrast of 'informal' with 'formal' suggests the contrast of lounge-suit with dress-uniform, and this was never the burden of the older 'formal'–'material' dichotomy; but it marks a greater readiness to acknowledge a fundamental difference than, for example, one could read into Whately.

Let us turn, then, to the second part of our task. Can analytical accounts of any particular Fallacies be found within the traditional and orthodox formal theories of inference? In the case of

[1] e.g. Copi, *Introduction*, ch. 3.

those which cannot, can any 'formal' analysis be found at all, or do they defy this kind of treatment?

The Fallacies that most clearly stand *outside* the scope of present-day logical systems are Equivocation, Amphiboly, and the other Fallacies Dependent on Language. Equivocal terms are excluded from formal systems by definition. Amphibolous formulae, or those liable to change of meaning through combination or division of terms, are regarded as not being well-formed. Some formulae are, no doubt, subject to misinterpretation after the manner of examples of Figure of Speech, but this is quite incidental to their formal features.

No attention whatever is usually paid to spoken forms as against written ones, as demanded in the Aristotelian analysis of Accent, and no formal role is attached to *emphasis*, as demanded in modern accounts. If it is objected that we could give metalinguistic or syntactical accounts of some of these – for example, of the amphiboly inherent in an expression such as '$2 + 3 \times 4$', in the absence of brackets or bracketing conventions – it may be replied that this is not a 'formal' analysis in the required sense; not, for example, the same thing as demonstrating, within a formal language, the existence of generic forms of Amphiboly or, among the rules of inference of that language, rules aimed at their avoidance.

It is, again, equally clear that present-day formal systems do not help us with analysis of the Lockean *ad hominem* and *ad verecundiam* arguments.

In the case of some of the Aristotelian Fallacies Outside Language, it is almost equally clear that a formal analysis is very much in order and, perhaps, a necessary part of any explication. The most obvious of these is Consequent which, as we have noticed already, was picked up and elaborated a little by J. N. Keynes.[1] Almost equally obvious is the classical Non-Cause as Cause, which we have had to analyse formally when describing it above in chapter 2: if assuming the truth of S leads us to an 'impossibility', or clear falsehood, we may deduce that S is false; but if, along the way in the argument from S to an 'impossibility', we make any other assumption T, the conclusion that S is false is invalid and must be replaced by 'Either S is false or T is false'.

[1] *Studies and Exercises*, pp. 353–4.

We can express the principle of the first, valid inference in the modern statement-calculus as

$$[(S \supset U) . -U] \supset -S$$

namely, as the *modus tollens*; and the second, invalid one as

$$\{[(S . T) \supset U] . -U\} \supset -S$$

which is easily shown by truth-tables not to be a theorem of the statement calculus. These formulae do not quite accurately embody the principles required, since the sign of material implication on the left-hand side does not quite accurately represent the concept of implication; but it may be replaced by a 'strict implication' sign within a modal calculus (any of the usual ones), or the whole formulae may be cast in, say, the 'derivation' logic of Popper ('New Foundations'), without destroying the fact that the first is valid and the second not. This is not the place to discuss subtleties in the concepts of Implication and Deduction: a formal analysis of Non-Cause as Cause is at least as adequate as these concepts will let it be.

Of the other Fallacies Outside Language, Accident, as we noticed in chapter 2, has sometimes been regarded, perhaps in the company of Consequent, as including all cases of invalid syllogism; and, if any formal analysis of it is to be given, it would be a modernized version of the theory of the syllogism. We also noticed, however, that some of the traditional examples introduce considerations outside these limits. One of these is the inference:

> You know Coriscus.
> You do not know the man approaching with his head covered.
> The man approaching with his head covered is Coriscus.
> Therefore, you both know and do not know the same person.

This could be regarded as an equivocation on the word 'know' on the grounds that it means 'are familiar with' in the first premiss and 'recognize' in the second; but alternatively we could regard it as a case of substitution in an 'opaque' context.[1] Let

---

[1] The recognition of the importance of these contexts is due, like so much in modern Logic, to Frege: 'On Sense and Reference' (1892). The term 'referentially opaque' is due to Quine, *From a Logical Point of View*, p. 142.

'K' represent a one-place predicate such that 'Kx' means 'You know x', and let 'c' be 'Coriscus' and 'a' be 'the man approaching'. The inference is now of the form

$$\frac{\begin{array}{c} Kc \\ -Ka \\ a = c \end{array}}{(\exists x)(Kx. -Kx)}$$

and the fallacy can be tracked down to the sub-inference

$$\frac{\begin{array}{c} -Ka \\ a = c \end{array}}{-Kc}$$

From the fact that I know someone under one description it does not follow that I know him under any other; witness Frege's example of the Morning Star and the Evening Star. What is necessary, if I am assumed to be a normally rational being, is that I should also *know* the two descriptions to be equivalent. Epistemic Logic, though it is usually formulated with a propositional epistemic operator rather than an epistemic predicate, can be axiomatized in such a way as to model this stipulation.[1]

The traditional Fallacies of *Secundum Quid*, and the Fallacies that Boethius and the medievals group under Misconception of Refutation, turn on the logic of various adverbial modifiers and the extent to which they are capable of changing the truth-values of sentences to which they are attached. Aristotle's leading example is that of the Ethiopian who is 'black' *tout court*, but 'white' in a certain respect, namely, in his teeth and eyeballs. The inclusion of an adverb such as 'wholly' or 'partly' in any sentence with a colour-predicate immediately clears up the confusion, and a formal logic of these adverbs is not difficult to build. Since, however, this logic has not been widely studied in modern times it is worth while to give a brief outline of it.

Adverbs of this kind tend to go together in 'squares of opposition' with logical properties resembling those of the traditional

[1] For example, an epistemic system is isomorphic to an S5 with one-place predicates and identity; preferably without the Barcan formula. See Hughes and Cresswell, *Introduction to Modal Logic*.

one. Let 'X-wise' be such an adverb: we frequently find that, for any individual A and any predicate P of some range, there are four statements related as follows:

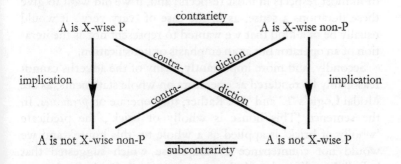

In its unnegated form 'X-wise' occurs at the top left-hand corner of this square; and we might call any adverb which has these properties a *top-corner* adverb. Examples of top-corner adverbs or adverbial phrases are: 'necessarily', 'absolutely', 'certainly', 'unconditionally', 'always', 'permanently', 'wholly', 'everywhere', 'compulsorily', 'in every respect', 'usually', 'in most places', 'in most respects', 'largely', 'probably', 'very'. Since the diagram has left–right symmetry there is no logical distinction between concepts of the top left-hand and top right-hand corners and we can add: 'never', 'nowhere', 'in no respect', 'seldom', 'in few places', 'hardly'. By contrast the following are *bottom-corner* adverbs or phrases: 'at least possibly', 'at least relatively', 'at least conditionally', 'at least partly', 'at least sometimes', 'at least somewhere', 'in at least a few respects'. (The words 'at least' are often, at some risk of misunderstanding, omitted.)

Since 'necessarily' is among the top-corner adverbs it would seem that modern Modal Logic is the appropriate vehicle for our formal theory. This is so, but there are two qualifications to be made. The first is that modern modal systems are generally far too rich: most of the adverbs make no sense iterated, as modal logicians iterate 'L' ('necessarily') and 'M' ('possibly') in formulations like 'LLp', 'MMp', '$LMLp$'. It is the differentiation of conceptions of the logical properties of these combinations that is largely responsible for the bewildering range of competing modal systems; and although (perhaps regrettably) the efforts of some of our best logicians have gone into this study, we cannot use many of their results. It makes very little sense to say that something is everywhere everywhere the case, or hardly hardly, or in most respects in most respects; and, if we did want to give these iterations a sense, as in the case of 'very very', it would usually be of a kind that we wanted to represent not as the iteration of an operator but as an emphasis or itensification.

Secondly, and more importantly, many of the adverbs cannot reasonably be rendered as operators on whole statements, as are Modal Logic's 'L' and 'M'. Rather, they operate on *predicates*. In the sentence 'This house is wholly of brick', the predicate 'wholly of brick' is applied as a whole to 'this house', and we would not countenance a paraphrase which suggested that 'wholly' modifies the whole sentence, as in 'It is wholly the case that this house is of brick'. Moreover, when the sentence is not of simple subject-predicate form but, say, a disjunction, 'wholly' does not make good sense at all: for example, in 'It is wholly the case that either this house is of brick or this fence is of wood'.

Top-corner adverbs do not need to be analysed all in the same way, and some of them can not only be meaningfully iterated but can also reasonably be regarded as sentential operators; but, for others, some new system-building is certainly necessary. There is no reason to suppose that this involves any very great difficulty. For illustration, let us continue with the properties of 'wholly'.

Let $a, b, c, \ldots$ be physical individuals and let 'f', 'g', $\ldots$ stand for colour and chemical-composition predicates such as can apply to parts of objects as well as to wholes: let 'f' mean 'is white'. Since to say of an object that it is white, or metallic, can be ambiguous unless it is specified whether it is wholly or only partly so, the predicates 'f', 'g', $\ldots$ are incomplete until prefixed by 'W' for 'wholly') or a similar operator. We can, however, form complex incomplete predicates out of simple ones by the use of negation, conjunction, and disjunction operators; and, in particular, we can formulate 'Wfa', '$W(-f)a$', '$-(Wfa)$', and '$-(W(-f)a)$' to mean 'a is wholly white', 'a is wholly non-white', 'a is not wholly white', and 'a is not wholly non-white'. The brackets in these formulae are not strictly necessary. Only a completed predicate, we shall say, may be applied to an individual-name. Formulae such as '$W(fvg)a$', for 'a is wholly white and/or black', are straightforward in meaning provided a little care is taken with the subtleties of 'and' and 'or'. The logic of the system is not very different from that of uniformly modal systems or the theory of one-place predicates.[1]

Aristotle's example of the Ethiopian can be regarded as turning on a confusion of 'a is wholly black' with 'a is partly black' ('Wfa' with '$-W-fa$'); or, rather more plausibly, a confusion of 'a is wholly black' with 'a is black' *tout court*, the latter meaning that a is mainly, generally or in principal respects black. Let us invent a special adverbial operator 'G' to mark this *tout court* sense: 'G' will generate its own square of opposition but 'Gfa' is implied by 'Wfa', though not vice versa. By this analysis the fallacy consists in supposing that 'The Ethiopian is black' ('Gfa') implies 'The Ethiopian is wholly black' ('Wfa') and hence that the Ethiopian is black in all individual parts.

The Boethian Fallacies of Different Part, Different Relatum, Different Time, and Different Modality can be similarly analysed: they arise from the use of incomplete predicates and vanish so soon as an appropriate adverb is inserted. 'The eye is white' and 'The eye is non-white' appear to be contradictories but are not so if they really mean 'The eye is partly white' and 'The eye is partly non-white'. 'Ten is double' and 'Ten is not double' – though no one, these days, would be even tempted to regard them as complete as they stand – are analogously reconciled by throwing in, in both cases, the adverb 'relatively'. 'Socrates is sitting down' and 'Socrates is not sitting down' are a little more difficult because of the special phenomenon of grammatical tense. Development of a logic of tenses is a special breakthrough of recent research,[2] and is usually based on sentential operators rather than operators on predicates, though there is a case for using the latter in rendering 'past Prime Ministers', 'future astronauts', and the like: the difficulty is that English and most other natural languages operate on a partly-different plan in providing, and demanding the use of, different tenses in their verbs. Logical

---

[1] Special axiom schemata: (1) $W\phi x \supset -W-\phi x$; (2) $[W\phi x . W(\phi \supset \psi)x] \supset W\psi x$. Special rule: $W\chi x$ for any tautological incomplete predicate 'χ'. Individual variables and quantifiers may be introduced in the usual way.

[2] See particularly Prior, *Past, Present and Future*.

tradition generally ignores this, and 'Socrates is sitting down' and 'Socrates is not sitting down' can be regarded as not-really-contradictory only if *either* (1) they are regarded as tenseless, and as meaning respectively 'Socrates is, was or will be at some time sitting down' and 'Socrates is, was or will be at some time *not* sitting down', in which case 'at some time' is our adverbial operator; *or* (2) they are regarded as present-tense but as uttered at different times. In the latter case ordinary Formal Logic can provide no analysis since context of use comes into the picture.

The final Boethian Fallacy involves the distinction between potentiality and actuality and this is regular modal territory. Even so the examples – 'The kitten can see' and 'The kitten cannot see' – seem better suited to an analysis in terms of operators on predicates than to one in terms of operators on whole sentences.

These analyses go as far as we could reasonably expect them to and, subject to some working out of detail, we can regard them as giving intrinsically satisfactory formal theories. It is possible to have reservations about them, however, as explications of the Fallacies concerned. These can be stated something as follows.

The purpose of the construction of a formal theory is, first, to bring out clearly features of our thought that ordinary language disguises and, secondly, to fill ordinary language's deficiencies. The invention of symbolic operators such as our 'W' and 'G' does not fulfil either of these purposes in a way relevant to fallacies *secundum quid*; because words like 'wholly' and 'mainly' already exist in English to do the job these operators do, yet fail to obviate fallacy because we choose not always to use them. The odious feature of our formal theory, considered as a remedial device, is its insistence that what we have called 'incomplete' predicates are deficient and need to be completed. When, in ordinary language, I describe an object as white, or as metallic, I should not be regarded as having made a deficient or ambiguous utterance. That there are certain respects in which it is *unspecific* is an essential characteristic of any utterance whatsoever; and 'There is a book on my desk' cannot be regarded as ambiguous simply because it fails to specify the book's colour.

Coupled with the tendency to regard '*a* is white' as ambiguous is the tendency to regard it as being capable of having this supposed defect cured; but adding an adverb such as 'wholly' or 'mainly' will not remove all question of unspecificity, because the judgement could still be either relative or absolute; and if this is resolved, it could still be either necessary or contingent, permanent or temporary, conditional on this or that. In short, there is no end to the further qualifications a statement might carry, and it is not the job of language to express them all every time. This being so, the Fallacy of *Secundum Quid* is an ever-present and unavoidable possibility in practical situations, and any formal system that avoids it can do so only at the expense of features essential to natural language.

This said, let us defer further consideration and turn to other cases.

We noticed that Ross credits Aristotle with the doctrine that a *question-begging* inference can be represented formally as a syllogism in which the conclusion follows from one of the premisses alone, independently of the other. This is certainly not Aristotle's view in the *Sophistical Refutations* but it may or may not have been his view later. There are obscurities in it but we are now in a position to give a precise account of at least one thing that might have been meant. The first question to be asked is: Should such a syllogism be regarded as *valid*? This might be answered by distinguishing, as we saw that Peter of Spain and J. N. Keynes did, between *inference* and *proof*. We might say that such a syllogism is inferentially valid, but invalid as a proof. All valid proofs are based on valid inferences, but not all valid inferences give valid proofs. The concept of 'proof', in this analysis, is being used in a sense which is appropriate only to Aristotelian-style deductive sciences, but we may postpone criticism on this point for the moment.

Now let us write 'Inf$(p; q)$' for 'There is a valid inference from p to q', and 'Inf$(p, q; r)$' for 'There is a valid inference from p and q to r', and so on; and similarly 'Pr$(p; q)$' and 'Pr$(p, q; r)$' for 'There is a valid proof from p to q' and 'There is valid proof from p and q to r'. These symbolic expressions specify various things about the logical relation between p and q, or the three-termed logical relation between p, q, and r. 'Inf$(p; q)$', in the first place, says no more than that p *implies* q; that is, p stands to q in the relation of superimplication or equivalence. In state-description-possibility terms, the state p is not possible, but other states, subject to the previously mentioned limitations, may or may not be. Similarly 'Inf$(p, q; r)$' specifies only the impossibility of the state $pq\bar{r}$.

'Pr$(p; q)$' is consistent with fewer state-descriptions than 'Inf $(p; q)$' since a statement q cannot be *proved* from a statement p equivalent to it, which would be at least as uncertain as it *a priori*: least of all can a statement be proved from itself. Hence 'Pr$(p; q)$' says that p stands in the relation of superimplication to q. State $p\bar{q}$ is impossible but state $\bar{p}q$ is possible; and so, for contingency of p and q, are pq and $\bar{p}\bar{q}$. 'Pr$(p, q; r)$' says that state $pq\bar{r}$ is not possible but, since r does not follow from p alone or from q alone, states $p\bar{r}$ and $q\bar{r}$, and hence states $p\bar{q}\bar{r}$ and $\bar{p}q\bar{r}$, must be possible. Further, r must not, perhaps, be *equivalent* to the conjunction of p and q, and p must not be inconsistent with q. There are various three-term relations consistent with this set of conditions,[1] and 'Pr$(p, q; r)$' can be regarded as stating that some one of these relations holds. Alternative, tighter or looser, definitions are possible but this one sufficiently represents the *genre*. Extension to multiple-premiss inferences and proofs is similarly straightforward in theory.

This analysis would also fit the Stoic Fallacy of Superfluous Premiss, and an argument with a superfluous premiss might be considered as inferentially valid but as not fulfilling the more demanding criteria of satisfactory proof. In fact, Superfluous Premiss and Begging the Question turn out to be more or less the same thing, and we might be led to wonder whether the genesis of Superfluous Premiss was not the currency, among the Stoics, of an explication of Aristotle along the same lines as that of Ross.

What merit does this analysis really have as a theory of Begging the Question? The distinction between the weaker and stronger relations, 'Inf' and 'Pr', is of some logical importance; but, as in previous discussions, we may have reservations about

---

[1] Extending the traditional doctrine of the seven logical relations to three-term relations, we find that there are 193 in all, of which 14 are consistent with the conditions imposed.

its relevence to our present inquiry. Even if we ignore the clearly dialectical force of the word 'beg', which is properly at home only where the argument in question has both a donor and a recipient, we may be dissatisfied with 'Pr' as an explication of valid proof. There is nothing wrong with equivalence of premiss and conclusion in a proof-process, provided the premiss is acceptable for reasons independent of the conclusion: I can prove either of 'Today is Tuesday' and 'Tomorrow will be Wednesday' from the other, provided only that the one is genuinely established and there is what nineteenth-century logicians used to call a 'movement of thought' accompanying the passage to the other, which is regarded as in need of proof. Nor is there anything in a superfluous premiss that need be considered to invalidate a proof. We shall take the conditions of proof up in earnest in the next chapter.

The Fallacy of Many Questions is given a virtually formal treatment by medieval writers or even by Aristotle. William of Sherwood traces the trouble to the negation of statements containing conjunctive nouns: 'Socrates and Plato are at home', he holds, implies straightforwardly that Socrates is at home and Plato is at home, but 'Socrates and Plato are not at home' does not imply either that Socrates is not at home, or that Plato is not at home, but only that they are not both so. In modern logical systems conjunctions of individual names are not used, because of some unwelcome complications they would bring; but, if we permit them temporarily for the sake of example, we could say that the Fallacy consists ultimately in treating

$$-f(a \cdot b)$$

as meaning

$$-fa \cdot -fb$$

whereas it should for consistency be taken as meaning

$$-(fa \cdot fb).$$

There is no mention, in this analysis, of *questions*; and, if we take the name of the Fallacy seriously we shall have to say that it consists in *asking a question* in a form such as 'Are Socrates and Plato at home?', in circumstances in which it is possible that one is at home and one is not. This takes us outside the scope of propositional Logic, and we need to look at very recent work to find a Logic of Questions. In connection with this work there has been a good deal of discussion of the Fallacy.

For a discussion that deal in depth with the Logic of Questions the reader must be referred elsewhere.[1] It will suffice for our purpose to consider a restricted class of questions, namely, those that can be represented as demanding choices between specified finite sets of alternative statements. Thus the question 'Did she wear the red hat, or the blue, or the white?' demands a choice between the three statements 'She wore the red hat', 'She wore the blue hat', and 'She wore the white hat'; the question 'Is John at home?' demands a choice between 'John is at home' and 'John is not at home'; and, apparently 'Has Jones stopped beating his wife?' demands a choice between 'Jones has stopped beating his wife', and 'Jones has not stopped beating his wife'.

The description of questions like the last one as *risky* is due to Belnap (p. 135). Let us symbolize the question which demands a choice between three mutually exclusive statements S, T, and U by '$?(S, T, U)$': this question is a *safe* one if S, T, and U are logically exhaustive in the sense that the disjunction $S \vee T \vee U$ is a tautology; *risky* otherwise. If A is 'Jones used to beat his wife' and B is 'Jones now beats his wife', the conjunction '$A.\text{-}B$' represents, without relevant inaccuracy, the statement that Jones has stopped beating his wife, and '$A.B$' represents the statement that he has not; whence the question is

$?(A.\text{-}B, A.B)$.

This is not 'safe': the disjunction $(A.\text{-}B) \vee (A.B)$ is equivalent to the question's presupposition, A.

Åqvist gives us three ways of making this kind of question 'safe'. One way is to treat the question as

$?(A.\text{-}B, -(A.\text{-}B))$

so that if Jones has never beaten his wife the correct answer to 'Has he stopped beating her?' is 'No'. Another way is to treat

---

[1] I recommend particularly: Belnap, *An Analysis of Questions: Preliminary Report*, and Åqvist, *A New Approach to the Logical Theory of Interrogatives*.

'He has never beaten her' as a proper third possible answer to the question, which becomes

$$?(A.-B,\ A.B,\ -A).$$

What Åqvist calls the 'Whately–Prior' method,[1] which treats the question as two separate ones, 'Used Jones to beat his wife?' and 'If he did, has he stopped?', is regarded by the Priors as different from this but by Åqvist as equivalent. Finally, Åqvist prefers to take seriously the concept of a *conditional* question and to give a logic of it; and consequently is able to give an analysis in terms of the second, alone, of Whately's and the Priors' two questions, namely, '*If* Jones used to beat his wife, has he stopped?', symbolised

$$?(A.-B,\ A.B\ /\ A)$$

where what follows the slash is the conditional clause. Briefly, the theory is that the question lapses if the specified condition is not satisfied.

On the Fallacy, however, Åqvist says (pp. 74–5):

> But I do not think that anyone would really contend that the Fallacy of Many Questions is committed by every risky question, i.e. by every question having some *possibly false* presupposition. The alleged fallacy is rather taken to be committed only by such risky questions as indeed have a *false* presupposition.

And these cases are more than just 'risky': they are cases in which the feared casualty has actually occurred. This is important since it indicates where the formal analysis stops. On a matter of detail we might object that it is not so much *false* presuppositions as *unwarranted* or *improper* ones that need to be singled out in practice for condemnation; but the point is that, in introducing these concepts – even the concept of falsity – we move outside the scope of Formal Logic. The scope of Formal Logic is enlarged when Questions are introduced into it, but it is still not so large as to include the contextual or dialectical concepts that are needed to give a fully-practical account of the Fallacy.

[1] From Whately's discussion of the Fallacy of Many Questions, in *Elements of Logic*, Bk. III, § 9; and Prior, M. L. and A. N., 'Erotetic Logic'. The latter article reopened the Logic of Questions after a long lull.

This might have been predicted. Presuppositions can occur as easily in statements as in questions and a Logic of Questions is not going to add anything to what a sufficiently perceptive logician could discover without it. The name 'Fallacy of Many Questions' has been often criticized and is misleading unless we see it purely in the context of Greek debates or the Obligation game.

Which others of the traditional Fallacies might it be profitable to try to analyse formally? There is precious little chance of a formal account of *ad hominem*, and not much of *ad verecundiam* or the other appeals to emotion. Arguments from authority, it is true, have been rather unfairly ignored by formal logicians: starting from the undoubtedly valid

> Everything X says is true.
> X said that P.
> Therefore, P.

we could be expected to find weaker, but still not 'fallacious', forms of argument within which some support is given to P by premisses of forms such as 'X is an authority on facts of type so-and-so'. Laplace's calculus of Testimony is probably useless; but *some* kind of calculus of Testimony must be of some use, however imponderable the factors in practical cases. Its construction will not be attempted here.

I shall finish this chapter with a sketch of a plausible suggestion for a formal theory of Fallacies Dependent on Language. It is not, in the event, very successful but, like some of the other proposals in this chapter, needs to be looked at if only to be disposed of. I shall call it the 'two-language theory'.

Equivocation – I shall use the word to include Amphiboly, which differs only in involving several words instead of one – is often regarded as due to imperfections of our language, which may use a single verbal formulation where there are two or more possible meanings. This condition is not necessarily remediable, since it is possible that, as Abelard says, there is a 'shortage of words'. Nevertheless it is possible to imagine that there might exist a perfect language within which no word or phrase had more than one meaning, and within which Equivocation could not occur. When we say that a word or phrase has more than one meaning we are, in effect, saying that it cannot be uniquely translated into this perfect language.

It is largely in the twentieth century that we have grown used to distinguishing regularly between 'propositions' and the 'sentences' within which they are formulated. There are other distinctions, too, which are sometimes of logical relevance: between sentence-types and sentence-tokens, for example, and between differently based arrangements of either of these into equivalent classes. We noticed that Aristotle's account of the Fallacies of Composition, Division and Accent presupposes a distinction between written language and spoken. However, at the risk of producing a theory that is uselessly naïve, let us ignore these other distinctions and concentrate on the dichotomy of sentence and proposition. (Much of what is said can, as it happens, be adapted to an account of Composition, Division, and Accent, if only because Aristotle thinks of written language as more fundamental or perfect than spoken, in the same way as 'propositions' are more fundamental than 'sentences'.)

Let us postulate, then, two languages to be known as the *S-language* (or sentence-language) and the *P-language* (proposition-language): and proceed to explore the thesis that Equivocation can be represented as a failure of unique translation between them. We can think of these, if we wish, as two actual languages, say Spanish and Portuguese. It is, no doubt, the case that there are sentences in Spanish that cannot be uniquely translated into Portuguese and vice versa. From the point of view of the speaker of Portuguese, some words in Spanish are equivocal and, from the point of view of the Spaniard the same is true of Portuguese. Conversely, Portuguese speakers find that Spaniards make some distinctions that are untranslatable; and, again, vice versa. Simple cases of both of these phenomena will occur to any student of language.

The Spanish–Portuguese analogy is, no doubt, inexact in a number of important respects; but particularly in respect of its apparent symmetry. In our own case we have an asymmetry built into the model, since we favour the P-language as the precise and logical one: we are professionally Portuguese. That this choice cannot be regarded as arbitrary is seen as soon as we start probing into the logic of the respective languages. What are the objects of logical relations: sentences, or propositions? Supporters could be found for both points of view. Thus Carnap:[1]

> ... the development of logic during the past ten years has shown clearly that it can only be studied with any degree of accuracy when it is based, not on judgments (thoughts, or the content of thoughts) but rather on linguistic expressions, of which sentences are the most important, because only for them is it possible to lay down sharply defined rules. And actually, in practice, every logician since Aristotle, in laying down rules, has dealt mainly with sentences.

Carnap's 'sentences', however, are sentences not of ordinary language but of an artificial language that aims at freedom from ambiguity, and this weakens the relevance of his stand. Logicians, of course, 'deal mainly with sentences', since this is the only way they have of expressing themselves; but when there is question of whether a sentence S *really implies* a sentence T, it is necessary to allow for possible ambiguities and hence to consider what propositions S and T convey. We must, as it were, start and finish in S-language; but we must look to P-language to provide our criteria of valid inference, and we must consequently translate our premisses to P-language and retranslate our conclusion.

The strength and weakness of actual languages, perhaps, is that they do part of our thinking for us. Most of the time, there is and need be no process of translation and retranslation, becase S-language and P-language march together. There are plausible 'syntactical' doctrines of implication applicable to S-language, which are capable of usually leading us to validity, as well as of occasionally leading us into fallacy. Our two-language model provides us with an explanation of why fallacious arguments seem valid, as well as of why they fail, in that it is possible to postulate a deductive system in *both* languages, that of P-language being the criterial one.

It is convenient, and sufficient for purposes of present illustration, to regard inferences as proceeding always from a single premiss to a single conclusion, or as consisting of chains of elementary inferences of this form. This assumption could be justified by postulating an 'Adjunction Principle', to the effect that

---

[1] *Logical Syntax of Language*, Introduction, p. 1.

every set of statements (of each language) has a *conjunction*, itself a statement of the language, such that assertion of the conjunction is equivalent to assertion of all members of the set. For short, the conjunction 'and' is available in both languages and unambiguous in all its dealings. Now, assuming that implication is defined in both languages and that there is a 'dictionary' relation associating sentences and propositions, we can start drawing diagrams illustrating various cases of interest.

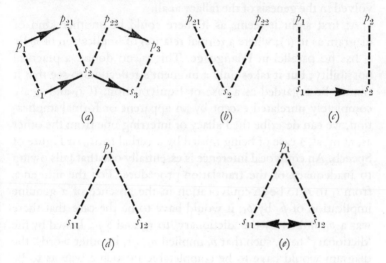

In figures (*a*)–(*e*) implication relations, whether in P-language or S-language, are illustrated by lines with arrows and 'dictionary'-translation is indicated by dotted lines. Figure (*a*) is the most obvious representation of an equivocal inference, in which s_3 is inferred from s_1, by way of an 'ambiguous' sentence s_2. The inference of s_2 from s_1 can be counted as valid since it is licensed by the more proper procedure s_1–p_2→p_{21}–s_2 and the same applies to the inference of s_3 from s_2; but, as a whole, the inference is faulty since there is no route in P-language from p_1 to p_3. On the assumption that the diagram is complete in relevant respects, there is no path from s_1 to s_3 that does not rely on the ambiguity of the intermediate s_2.

The curious part of this explanation is that we seem to be committed to the thesis that it may be *valid* to infer s_2 from s_1, and *valid* to infer s_3 from s_2, without being valid to conduct the two inferences in sequence. Validity of inference, between sentences, is not transitive. Perhaps, after all, it is only sentence-meanings, and not sentences themselves, that can figure in inferences. If this is so, most of figure (*a*) is irrelevant to the fallacious inference and the essential part of it is just as in figure (*b*). The move from s_1 to s_2, after all, is valid only so long as s_2 is taken in the sense p_{21}, and not in the sense p_{22}. Consequently the implications, whether of P-language or S-language, are not involved in the genesis of the fallacy at all.

At first sight it seems as if there could be another kind of diagram as in (*c*); where a formal relation of implication of s_2 by s_1 has no parallel in P-language. This is, no doubt, a practical possibility; but it takes only a moment's reflection to see that it cannot be regarded as a case of Equivocation. If s_1 and s_2 are completely unrelated except by an apparent or formal implication, we can describe the Fallacy of inferring one from the other as, at most, a case of being misled by a verbal form, or Figure of Speech. An equivocal inference is essentially one that fails owing to inadequacy of the translation-procedure. For the inference from s_1 to s_2 to be an equivocation in the absence of a genuine implication of p_2 by p_1, it would have to be the case that there was a p_i, linked by the 'dictionary' to s_1, and a p_j, linked by the 'dictionary' to s_2, such that p_i implied p_j; or, in other words, the diagram would have to be completable in such a way as to be equivalent to part of figure (*a*), and to contain a section as in figure (*b*).

Figure (*d*), a converse of (*b*), represents the case of synonymous sentences which might appear to be different in meaning because of their lack of formal relation. This case is of little interest since, it seems, we generally readily recognize linguistic equivalences with no formal basis, and since, in any case, fallacious arguments cannot be based directly on failure to do so. So soon as the equivalence is recognized, (*d*) is better represented as (*e*).

Let us return, then, to figure (*b*), as the basic representation of Equivocation, supplemented with (*c*) as the basic representation of Figure of Speech. It should now be clear that the apparatus of parallel languages and parallel implications has not, in fact, thrown much light on these Fallacies. When we think, as we must, of P-language as the fundamental one, the only influence the presence of S-language has is to throw in certain spurious logical relations: in case (*b*), a spurious equivalence between p_{21} and p_{22}, and in case (*c*), a spurious implication of p_2 by p_1. We would do just as well to model Equivocation and Figure of Speech on a single language, in terms of a supplementary web of apparent logical relations. We might, for example, consider the effect of having an 'apparent implication' operator in addition to a 'real' one. There will be, in all, three kinds of inference in such a system: 'real' inferences, purely 'apparent' inferences, and those inferences that result when both kinds of implications are allowed. There are various possible suppositions about the properties of the two kinds of operator: they may be consistent or inconsistent, 'apparent implication' may or may not be truth-preserving, and so on.

The fault of this model is that there is nothing in it that makes it specifically a model for Fallacies Dependent on Language. Every kind of fallacy whatever arises from an 'apparent implication'. The 'two-languages' model, which led us in turn to this one, has failed to provide the analysis for which it was designed. What we wanted was an analysis not of apparent implication in general, but of apparent implication due to multiple meanings of linguistic expressions. Multiple meaning is evidently not illuminated by the supposition that 'meanings' are expressions in an alternative language.

We shall later be led to analyse Equivocation, and the related Fallacies, very differently from this. The nominalism of Carnap finds its ultimate expression not in the erection of a distinction between words and their meanings, but rather in some dissolution of the concept of meaning into that of systematic use; and, if 'meaning' goes, 'equivocation', which is variability of meaning, will have to go too. We must postpone this discussion, however, until we have built some groundwork for it.

CHAPTER 7

The Concept of Argument

A fallacy is a fallacious *argument*. Someone who merely makes false statements, however absurd, is innocent of fallacy unless the statements constitute or express an argument. In one of its ordinary uses, of course, the word 'fallacy' means little more than 'false belief'; but this use does not concern us. In logical tradition, a fallacy may be made up even out of true statements, if they occur in proper form; that is, if they constitute or express an argument that seems valid but is not.

The concept of an *argument* is quite basic to Logic but seldom examined. However, there are problems which require us to take a closer look at it. We have felt the advance rumblings of several of these already. I shall be concerned principally with three of them.

(1) First, consider the problem of 'nailing' a fallacy. In many cases of supposed fallacy it is possible for the alleged perpetrator to protest, with an innocent face, that he cannot be convicted because he has not been arguing at all. Consider the so-called *argumentum ad hominem*, in the sense of the modern books. Person A makes statement S: person B says 'It was C who told you that, and I happen to know that his mother-in-law is living in sin with a Russian': A objects, 'The falsity of S does not follow from any facts about the morals of C's mother-in-law: that is an *argumentum ad hominem*': B may reply 'I did not claim that it followed. I simply made a remark about incidentals of the statement's history. Draw what conclusion *you* like. If the cap fits...' This would be disingenuous, but the point remains that B cannot be convicted of

fallacy until he can have an *argument* pinned on him. And what are the criteria of that?

To take another case, which is in earnest in that it is of a kind that can occur even in the relatively rarified atmosphere of philosophical discussion, consider the kind of move that we might be tempted to classify as argument in a circle, or question-begging. X asserts that the Principle of Non-Contradiction, 'A thing cannot have property P and property non-P at the same time and in the same respect', must be a valid principle and says 'If it were not, the world would be incoherent in that, for example, the chair I am sitting in might be existent and non-existent at the same time, or non-solid at the same time as it is solid'. But to show that such a world would be *incoherent* he must invoke the Principle of Non-Contradiction, which it is the object of the exercise to prove. Pressed, he says 'What I said is not circular because it is not an argument. I am not saying that the incoherence of a world containing contradictions is an *argument* for the Principle of Non-Contradiction. I am simply exemplifying the Principle in order to make clear what you could be committed to if you were to drop it'. So long as he can insist that he is merely *elucidating* his position, and not arguing, he can evade censure. If he is to be effectively answered, it must be on the grounds that what he says does constitute an argument, and based on an appreciation of what this involves.

(2) Secondly, consider the problems surrounding arguments on the fringe of Formal Logic: inductive arguments, arguments from authority. Is there such a thing as inductive validity, or is it a contradiction in terms? Although we accept in principle that some inductive arguments are better than others, what are the canons by which we judge an inductive argument's absolute, rather than relative, worth?

In the case of arguments from authority – since we cannot abjure these arguments altogether – how are we to balance authorities against one another, and arguments based on their opinions against arguments, such as inductive ones, from other sources? Is it always unreasonable to use an argument from authority against a deductive one (for example, in Mathematics)?

A prior question, both in the case of inductive arguments and in the case of arguments from authority, is: *Are they really arguments?* The logician commonly conceives arguments on the pattern '*P*, therefore *Q*'; but neither of these kinds of so-called argument fits easily into this mould. We do not normally say 'This crow is black; that crow is black; therefore, all crows are black', or 'The dealer said it's genuine Louis; therefore it's genuine Louis'. Instead, we frame, at most, a modified conclusion, in the form 'Therefore it is a reasonable conclusion that...', or 'So probably...', or 'So presumably...'. To call these 'arguments' is to mark a similarity to deductive arguments; but it might be as well to reassure ourselves that the similarities are really as great as the differences.

(3) A third problem involving re-examination of the concept of argument is the one raised by the assertion of Sextus Empiricus and J. S. Mill that every valid argument is question-begging. Mill says (*System of Logic*, Book II, ch. 3, §2):

> It must be granted that in every syllogism, considered as an argument to prove the conclusion, there is a *petitio principii*. When we say,
>
> > All men are mortal,
> > Socrates is a man, therefore
> > Socrates is mortal;
>
> it is unanswerably urged by the adversaries of the syllogistic theory, that the proposition, Socrates is mortal, is presupposed in the more general assumption, All men are mortal...

The key phrase for our purpose is 'considered as an argument to prove the conclusion'. Mill's position is clear and very simple: to get to any conclusion you must start from some premiss or premisses and, if the argument is valid, the premisses must be at least as strong as the conclusion so that, in assuming the truth of the premisses you have already assumed the truth of the conclusion (Book II, ch. 3, §3):

> The error committed is, I conceive, that of overlooking the distinction between two parts of the process of philosophising, the inferring part, and the registering part, and ascribing to the latter the functions of the former.

But this is to take a view of *argument* as an extended process that includes the verification of the premises on which inference proceeds. Mill is content to raise no question about the verification of singular propositions such as 'Socrates is a man' but demands of an argument, in effect, that it contain no universal premisses of the form of 'All men are mortal' since these necessarily indicate that it is incomplete. He goes on to develop the theme that the true process of reasoning is by what has sometimes been called 'analogy', from particulars directly to other particulars.

A presupposition of Mill's doctrine is that the primary and only true purpose of argument is to establish 'scientific' knowledge. But what of other contexts? De Morgan complains (*Formal Logic*, pp. 296–7):

> It is the habit of many to treat an advanced proposition as a begging of the question the moment they see that, if established, it would establish the question.

He does not tell us what to do about this; and, if what Mill says is to be accepted, it is difficult to see that the behaviour referred to is not, usually, completely proper.

It does not occur to Mill, as it does to De Morgan, that there are dialectical criteria bound up in the notion of question-begging. Even a Millian argument from particulars to particulars could be open to the charge of question-begging on these other criteria. Thus someone who argues that democracy will be unsuccessful in New Guinea because it has been unsuccessful in Ghana, Ceylon, and Vietnam could have it conceded that his analogical inference process is a valid one but be charged with begging the question in his premisses.

Mill's doctrine has been important in helping to create a style of thought about philosophy, which is characteristically modern in its consequences. Philosophers grasp the proferred nettle and say 'Philosophical arguments, above all others, are circular really. A philosophical argument, having no empirical foundation, never teaches anyone anything he doesn't know already'. Yet they defend the so-called 'teaching' of Philosophy. They cannot have it both ways. If philosophical arguments really lead nowhere they should be dropped, and philosophers should stop drawing their pay. But perhaps, of course, there is more to be said about what an argument really is.

I think that, if we give an accurate account of what an argument is, we completely dispose of this third problem, and go a long way towards drawing the sting from the other two. Moreover, we lay a foundation for an understanding of the fallacy-tradition and its place in the study of Logic.

An argument is generally regarded as being whatever it is that is typically expressed by the form of words 'P, therefore Q', 'P, and so Q', 'P, hence Q'; or, perhaps, 'Q, since P', 'Q, because P'. As a brief run-down on the appropriate terminology, let me quote from Whately (*Elements of Logic*, Bk. II, ch. III, § 1):

> Every argument consists of two parts; that which *is proved*; and that *by means of which* it is proved. The former is called, *before* it is proved, the *question*; *when proved*, the *conclusion* (or *inference*); that which is used to prove it, if stated *last* (as is often done in *common discourse*) is called the *reason*, and is introduced by '*because*,' or some other *causal* conjunction; e.g. 'Caesar deserved death, *because* he was a tyrant, and all tyrants deserve death.' If the Conclusion be stated *last* (which is the strict *logical form*, to which all Reasoning may be reduced) then, that which is employed to prove it is called the *premises*, and the Conclusion is then introduced by some *illative* conjunction, as 'therefore,' e.g.
>
> 'All tyrants deserve death:
> Caesar was a tyrant;
> *therefore* he deserved death.'

We might reasonably ask why Whately should refer to 'the strict logical form' and why, in fact, anyone should consider the precise order of the premisses and conclusion to be of any logical importance. One partial, though inconclusive, answer might be found in the comparative ambiguity of 'because', which may herald either a causal (in the natural scientific sense) or a rational explanation, either a fact allegedly related to the already-stated fact as cause to effect or a statement which is alleged to be related to the previous statement in accordance with the canons of Logic.[1] A formulation with 'therefore' might, for this reason, be preferred as less ambiguous. Perhaps, however, the attitude is merely con-

[1] cf. Ryle, 'If, so and because'.

ventional and no more than another example of how tied we still are to Aristotle's apron-strings. We have already noticed that the Indian tradition has even more elaborate predilections. We must dispense with this kind of bureaucracy.

When we divide the statements making up an argument into *premisses* and *conclusion* we are importing another fixed idea; for many arguments in practice have a 'thread', a 'development' that involves intermediate statements belonging to neither of these categories. It is usually assumed in logic books that a complex argument can always be broken down into simple steps in such a way that, in any given step, there are one or more premisses, just one conclusion and no intermediate statements. This is true of some arguments but not of all; and the word 'argument' is, in any case, regularly and properly used of the complex of steps as well as of the steps themselves. If we do not bear this in mind we are tempted to give too simple an account of various important logical phenomena. For example, 'circular' arguments may be quite misrepresented if we treat them as one-step events.

On the other hand, an argument is more than just a collection of statements. 'P, therefore Q' states P and states Q, but there are other ways of stating P and Q that do not amount to arguing from P as a premiss to Q as a conclusion. You can say 'P, and moreover Q', indicating that P and Q are true but that Q goes further than P; or 'P, nevertheless Q', indicating that P and Q are true but that you wouldn't expect to find them true together; or you can just say P and then say Q. When you choose to say 'P, therefore Q', the important feature of your utterance is that, as well as stating P and stating Q, you *adduce P in support of Q*.

Now it is important to notice that when P is adduced in support of Q, it may actually *not* support Q. This is only to say that an argument may be invalid. However, it is important to emphasize that an argument is not to be identified with an implication. There may be an argument where there is *no* implication: I may argue from P to Q when P does not, in fact, imply Q. 'Argument' is not synonymous with 'valid argument'. Although the existence, in some sense, of a valid implication may be a necessary condition of a *valid* argument it is not a necessary condition of an argument. Conversely, that P would support Q if it were adduced to do so does not, in itself, imply that, when both are stated, an argument is in process. A person may state P and then particularize it to Q, or state Q and then go on to make the stronger statement P, without having argued in either case. He may even, in stating both P and Q, imply that it is Q that represents an argument for P instead of the other way round; or that P is an argument *against* Q and that therefore his joint statement poses a paradox; or any one of a number of other things. The forms 'P, and therefore Q', 'P, and moreover Q', 'P, and in particular Q', 'P, but nevertheless Q' are alike in that they all represent statements of both P and Q and imply a relation between them; but the actual relation between P and Q may not be that implied, just as P and Q may fail to be actually true as affirmed.

The actual logical relation between premisses and conclusion of an argument may be anything at all. It is even possible to find plausible arguments such that the conclusion is the precise contradictory of the premiss. People have argued[1]:

> Every event has a cause.
> Therefore (if you trace the causal sequence back far enough) some event has no cause.

We would not regard this argument as a valid one, and the fact that the conclusion contradicts a premiss is good prima facie evidence that it is not. Even so, there are arguments which could conceivably be regarded as valid in spite of this little failing. Consider this variant of the 'liar' paradox:

> Epimenides was telling the truth when he said 'I am lying'.
> Therefore, Epimenides was lying when he said 'I am lying'.

We can, if we choose, hold firm to the conviction that an argument *cannot* be valid if the conclusion contradicts a premiss; and, if we do, we are forced to find a fault in the reasoning in this example, such as by insisting that 'I am lying' is not a genuine statement. In place of this rigid attitude, however, it would seem better to admit that there are circumstances within which accepted inference-processes may lead to unacceptable conclu-

---

[1] I owe the example to D. C. Stove, who gave it in a paper read in Sydney in 1964.

sions and that, if we have to, we can learn to live with this situation: the acceptability of an inference-process is not a knock-down guarantee of the results to be obtained by its use, and arguments may have counter-arguments. Another example, of importance in twentieth-century logical history, is:

> No class is a member of itself.
> Therefore (since it follows that the class of classes that are not members of themselves is not a member of itself, and from this that the class of classes that are not members of themselves *is* a member of itself), at least one class is a member of itself.

This is not at all obviously invalid. We shall only insist that it is invalid if we think, on principle, that such an argument *must* be invalid. In fact, we need to make quite complicated and unwelcome revisions of logical system to accommodate the thesis that it is invalid.

I am not suggesting that logicians should accept defeat and abandon their quest for a paradox-free theory of deduction. The point is just that, whatever the result of that quest, there are various criteria of worth of arguments; that they may conflict, and that arguments may conflict; that when criteria conflict some are more dispensable than others, and that when arguments conflict a decision needs to be made to give weight to one rather than another. All this sets the theory of *arguments* apart from Formal Logic and gives it an additional dimension. This should now be abundantly clear, and we may turn our attention to the filling-in of details.

There is little to be gained by making a frontal assault on the question of what an argument *is*. Instead, let us approach it indirectly by discussing how arguments are appraised and evaluated. I shall stop using the words 'valid' and 'invalid' in case they cause concentration on too narrow a feature of this process of appraisal: the question interests in its broadest, rather than its narrowest, aspects. To avoid jargon as much as possible let *good* arguments be described simply as 'good'. There is no obvious or straightforward way of characterizing arguments that fall short of being 'good', because there are different ways of falling short; but I shall try to use the correct everyday word when there is one.

What are the criteria by which arguments are appraised?

The first thing we need to do is to deny one thing that most of the elementary logic books affirm. A distinction is faithfully made between the truth or falsity of the premisses and conclusion, on the one hand, and the validity or otherwise of the inference process on the other. A valid argument, it is said, may be built on completely false premisses and it may thus have a completely false conclusion. But this is a complete misrepresentation of the nature of argument. Take an example: I am arguing that the Viet Cong will be unable to last another year. I say 'American troops wear bright red uniforms. Bright uniforms excite envy in an enemy. Envious soldiers fall an easy prey to hardening of the arteries'. Given a couple of obvious additional premisses these imply the conclusion and those who follow the textbooks will have to say 'Well, it's a good *argument* . . . The conclusion is well supported by the premisses, and the fact that the premisses are not all they might be has nothing to do with the case'. This is obvious nonsense: whatever the textbooks say, *in practice* we like our premisses to be true, and we do not describe an argument as a good one if the premisses are false.

What about the conclusion? Presumably, if a good argument has true premisses and a satisfactory inference-process it must have a true conclusion too? Unfortunately the case is not quite so simple as this. If logicians had found their perfect theory of deductive validity and we were to agree to work within the bounds of this theory, this would, of course, be so; and, in time, we might become so sure of our theory that we come to regard it as a simple tautology that it is so. But this is not the case at present, and may never be; and, in any case, there are good arguments that are not deductive. In practice, although we would want to say of a good argument that it *supports* its conclusion, it is not, as a rule, possible to say that it supports it beyond fear of reproach or criticism. It often occurs that there are good arguments *for* a given conclusion and also good arguments against it. We cannot demand of an argument that it be, all by itself, a knock-down one. If we did, we would risk running across a situation in which we found that there existed both a knock-down argument for a conclusion and a knock-down argument against it at the same time. It follows that our proposed stipulation that the conclusion of a good argument must be true cannot be sustained unqualified.

I shall enlarge on this in due course. For the moment, let us take some time off to answer an objection.

It will be said: Arguments occur not only in the form 'P, therefore Q' or 'Q, because P' but also, sometimes, when we discuss the *passage from* the premiss to the conclusion, without being committed to the premiss or the conclusion themselves. We say '*If* P, *then* Q'; and in this form an argument can be presented, discussed, validated and agreed to quite independently of whether P and Q are true or false. In some sense, in fact (it would be said), this is the proper form of an argument so far as the logician is concerned, because he is not involved in the question of the actual truth or falsity of the statements in his examples, but only with the inference-process that they exemplify. It must be added that good or valid inference-processes are good or valid all by themselves, independently of the material to which they are applied.

The answer to this is that 'If P, then Q' is not a real argument at all, but only *a hypothetical* argument. It says that a certain hypothetical statement P, which I am not now making, would serve, *if* I were to invoke it, as a premiss for a possible conclusion Q; but the argument remains hypothetical because I do not, or not necessarily, now argue in this way. A real argument has real premisses and conclusion, not hypothetical ones.

To those accustomed to logic's traditional terminology this use of the word 'hypothetical' looks like a pun, confusing (*a*) an argument which is hypothetical in the sense of not being real or actual, with (*b*) an argument which is hypothetical in grammatical form, having a clause introduced by 'if'. If we think this through, however, we may come to wonder whether the senses are really as different as they seem. If someone wishes to hypothesize an argument the natural way to do it is to use a 'hypothetical' form of words, and our reason for using the description 'hypothetical' for this relevant form of words is precisely that it is the standard method of representation of an argument that is only hypothesized, not used, as it were, in anger. Over the centuries the word has been detached from some of its proper associations and become an only semi-meaningful piece of logicians' jargon, often misused in that even actual arguments will sometimes be described as 'hypothetical' if they contain a premiss hypothetical in form.

That this terminological muddle is part of the wall which shuts reality out of much of our theory can be seen when we reflect that examples in logic books are mostly hypothetical ones anyway, even when they are in 'therefore' form. When logicians confine themselves to examples of the form 'If P, then Q', they are consequently confining themselves to *hypothetical hypothetical* cases. At the very least, we should move one step closer to reality. When we put up an example of an argument we should imagine someone actually arguing, not merely imagine someone imagining someone arguing. It is very easy, later, to ascend the theoretical ladder by conditionalizing what is said; but it is not nearly so easy, if we start from the other end, to restore the additional dimensions of actuality.

Now let us return to the task of formulating criteria of appraisal, and start to put them down systematically. This is, at first, only a first attempt, and we shall soon find reason to make some mendments. The first two criteria, however, are fairly obvious:

(1) *The premisses must be true.*
(2) *The conclusion must be implied by them* (*in some suitable sense of the word 'implied'*).

Implication may be strong or weak, and the argument strong or weak accordingly. It is not here to be interpreted with the canons of any particular formal system in mind, least of all an exclusively deductive system. There is, however, no synoptic theory of implication in existence and we shall have to leave the concept in this vague state for the moment.

Let us go on to another requirement. It is not enough that the conclusion follow from the premisses: it must follow reasonably immediately. Take some fairly advanced theorem of Geometry; say, that the opposite angles of a cyclic quadrilateral are supplementary. Although this follows from any suitably complete set of geometrical axioms, it would not be sufficient, by way of premisses for this conclusion, just to give such a set of axioms. The axioms are the *starting-point* of the argument, but the argument itself has not been given until it has been spelt out. So we shall have to stipulate:

(3) *The conclusion must follow reasonably immediately.*

In practice, a complex argument resolves itself into a chain of simple arguments; and one of the objections to an argument that is not spelt out is that it is not clear in which of various alternative ways it is to be broken down. However, this is not the only objection: there may be only one way of breaking an argument down, and yet it may be criticized simply on the grounds that it is incompletely stated. It seems to be a part of the conception of an argument that it is in principle capable of being spelt out completely and in this case each step is such that the intermediate conclusion is reasonably immediately inferable from the, perhaps intermediate, premisses.

However, the premisses of an argument are frequently not given in full and we need a supplementary criterion to deal with this possibility. There are rules or, at least, conventions governing what may be omitted. If I were to say, using Whately's example

Caesar deserved death because he was a tyrant.

you would quite cheerfully supply the missing premiss

All tyrants deserve death.

and, in general, when the argument is in the form of a traditional syllogism, either premiss may be omitted. In other cases there can be doubt what premiss should be supplied and, of course, the recipient of the argument may supply the wrong one, or even supply none and unfairly declare the argument faulty. Without going into these details, let us just note our supplementary criterion shortly:

(4) *If some of the premisses are unstated they must be of a specified omissible kind.*

If the conclusion of an argument follows reasonably immediately from true stated premisses, or from these together with other true premisses of an omissible kind, or may be broken down into a chain or network of sub-arguments each satisfying these conditions, I shall say that it passes the *alethic* tests of goodness or worth. (The word 'alethic' is von Wright's word,[1] from

[1] *An Essay in Modal Logic*, p. 1.

the Greek word for 'truth'.) Arguments which pass these alethic tests can be regarded as setting a certain theoretical standard of worth, corresponding with a certain conception of 'pure' Logic.

We may or may not be concerned with 'pure' Logic in this book, but we are certainly concerned with the Logic of practice. Consequently it is important to move on to the additional or modified criteria of appraisal that are relevant when Logic is put to work. It would not be acknowledged by everyone, of course, that the criteria given are in any need of alteration. It would be convenient if this were so. Unfortunately, there is one very important respect in which the alethic tests are not sufficient, and another important respect in which they are not necessary.

Let me first take up the respect in which they are not sufficient. By the alethic tests an argument is a good one if the premisses are true and the conclusion immediately follows from it. But what is the use of an argument with true premisses if no one *knows* whether they are true or not. If I argue that the Martian canals are not man-made because there never has been organic life on Mars, or that Australian aboriginal culture is related to European because there was extensive prehistoric migration from Assyria, my premisses may be true but the arguments will be quite useless in establishing my conclusions so long as no one knows them to be true. And the argument that oranges are good for orang-utans because they contain dietary supplements might or might not carry some weight in the second half of the twentieth century but would rightly carry none at all as between two ancient Romans who had never heard of vitamins. The recipient of an argument of this kind will rightly challenge it with 'How do you know?'; but this attacks not so much the *truth* of the statement as its *epistemic* status. It is not enough for the premisses of an argument to be true: they must also be known to be so, and we must replace the alethic rule (1) by a corresponding epistemic one

(E1) *The premisses must be known to be true.*

Since whatever is known to be true must be, in fact, true this rule is stronger than the one it replaces.

Furthermore, a similar point applies to the inference-process. We generally suppose that all inferences that are not complex ones are public and obvious, and this may be true of most; but it is possible in principle that Q should follow from P but that the step be an obscure or logically tricky one that would not be generally recognized or acknowledged. There is, perhaps, something philosophically repugnant about the suggestion that there should be inference-processes that *nobody* recognizes; but there is nothing in the least odd in supposing that someone or some group of people should, say, have immense difficulty with *modus tollens* or *reductio ad impossibile*. Such people would have to prefer other argument-forms.

We tried to rule this situation out earlier by specifying that inferences must be 'reasonably immediate'; but this phrase may have alternatively an alethic or an epistemic interpretation, and it is now clear that, in practice, it is the epistemic one that really matters. However immediately Q may follow from P in an alethic, or logical-systematic, sense, the argument from P to Q is less than inadequate if it is apt to strike people as obscure. It needs to be replaced by one which is not merely immediate but acknowledged to be so. In an epistemic strengthening of the alethic rules (2) and (3) we shall be able to combine the two:

(E2, 3) *The conclusion must follow clearly from the premisses.*

It should be emphasized, still, that the inference need not be 'deductive' in the sense of being sanctioned by any particular logical calculus: it may be inductive, or extrinsic, or of a form for which no calculus has been developed.

A similar change of interpretation will now attend the rule about dropped premisses: 'omissible' can be regarded as explicated in some formal, non-epistemic way or it can alternatively be regarded as meaning 'acknowledged as omissible'. We can, in fact, make a contribution towards explaining what it is for premisses to be omissible: any premiss may be omitted that is known to be true and will be taken for granted in the context. That a premiss is 'known to be true' is not quite enough to licence its omission, since this would not distinguish it from premisses that need to be stated; and we frequently need to be reminded, in argument, of things that we know. However, if 'being taken for granted' is a concept admissible into epistemic Logic, we may state our next rule:

(E4) *Premisses that are not stated must be such that they are taken for granted.*

They must, of course, also be known to be true, by (E1).

Whether the epistemic appraisal-criteria, taken together, are stronger than the alethic ones will depend on details of interpretation, such as whether 'taken for granted' in (E4) is really stronger than 'omissible' – which must be regarded as an alethic term – in (4). In the case of the epistemic criteria, however, there is an additional one; for an argument is surely unnecessary and dispensible unless its conclusion is *not* known to be true – that is, is in doubt – before the argument is put forward. It is true that there are 'academic' contexts, as we say, in which we produce or run over new arguments for old conclusions that are already well supported; but these, again, are *hypothetical* arguments or, at best, rehearsals for actual ones to be carried out on other victims on other occasions. It is as if we said: '*If Q* were not already known to be true it could be supported as follows: *P*, therefore ...'. Our rule is:

(E5) *The conclusion must be such that, in the absence of the argument, it would be in doubt.*

This brings us to a difficulty. When epistemic logics have been formalized it has been usual to treat the 'knowledge' that is expressed in the symbolic operators as if the fictitious 'knower' were a person of infinite logical wisdom and rationality. It is implicit in the axioms and rules of the system that all logical theorems are 'known' to this person and that He draws all logical conclusions from whatever He knows. If 'K' is an operator meaning 'It is known that' it will commonly be regarded as a rule of the system that

$\quad$ *If α is a theorem so is $K\alpha$*

and among the theorems we shall have

$$[Kp \cdot K(p \supset q)] \supset Kq.$$

The assumptions, however, tend to make nonsense of our epistemic rules. If P really implies Q and if P is *known* (to this person of perfect logical wisdom), then Q will already be *known* also; and no argument will ever have any effect, since its conclusion will already be known as soon as the premisses are. Another way of putting this would be to say that when, in stating rule (E5), we said that the conclusion must be in doubt *in the absence of the argument*, we did not make it at all clear what difference the actual *stating* of an argument makes to the logic of the situation in which it is stated. If we are to treat epistemic concepts like 'known' and 'in doubt' as logical ones, it might be said that we cannot allow the truth of statements involving them or the validity of logical relations formulated in terms of them to be contingent on the historical accident that someone actually *says* something or other on a particular occasion.

The answer is that epistemic concepts do not have to be logical concepts in this puristic sense for it to be worth our while to express them in symbols; and that the axioms and rules referred to do not satisfactorily express the logic of 'It is known that', when the knowers concerned are of less than perfect logical wisdom. Dropping these axioms and rules and replacing them with more reasonable ones does not present any very formidable difficulty; and it permits us to preserve rule (E5). One further comment should be made. The terms 'known', 'in doubt' and 'taken for granted' have been used as if there were no distinction to be made between different knowing subjects, so that what is known to one is known to all; but they are really doing duty for a range of concepts, 'known to me', 'known to you', 'known to John Smith', 'known to modern Science', 'known to most members of the diplomatic establishment', and so on. This does not make much difference so long as we stick systematically to one of relevant contexts. Thus, if the arguments we are discussing are arguments that John Smith produces within his own head and for his own edification the appraisal-criteria will refer exclusively to what is known to John Smith, in doubt to John Smith, and so on. However, the paradigm case of an argument is that in which it is produced *by* one person *to convince* another. Generally, the concepts relevant are those that refer to the person the argument is aimed to convince; but we can imagine complications in case, say, the arguer wishes to argue that the other person 'should know' such-and-such, or onlookers try to evaluate the outcome in *their* own terms. These make us wish for the simplicity, unfortunately illusory, of the alethic case. We shall see, however, that the task of building a formal model of some of the epistemic phenomena referred to is not unbearably difficult.

Before moving on to a further modification of the criteria it is worth remarking that one tempting generalization of them, the introduction of *degrees* of knowledge and doubt – that is, epistemic probability – turns out, when an attempt is made to formulate it in detail, to be much less clear than it seems at first. We feel that it should be possible to weaken (E1) to

(P1) *The premisses must be reasonably probable*

and (E5) to

(P5) *The conclusion must be less probable* a priori *than the premisses.*

It is less clear that there is any appropriate relevant way of altering the other rules. If the premisses clearly imply the conclusion it would seem that they would be capable of raising the *a priori* probability of the conclusion to their own figure; but it would be equally sensible to suppose that, since the negation of the conclusion implies the negation of the (conjunction of) premisses, the confrontation of premisses with conclusion would operate the other way and reduce the probability of the premisses. A probabilistic argument, in fact, never seems to work very well unless the premisses are in some sense firmer and less open to revision than the conclusion is. In practice, no one is going to be much interested in a probabilistic argument unless the probability of the premisses very clearly outweighs the *a priori* improbability of the conclusion. The probability of the premisses, in short, must be fairly high, and the independently determined probability of the conclusion not very low.

Now let us consider the respect in which the alethic criteria are too strong. Since the epistemic criteria are, on most counts, stronger than the alethic ones, it will follow that the epistemic ones are too strong also.

The point concerns the strong connotations of the word 'know'. We felt the need to alter criterion (1), which says that the premisses must be *true*, to (E1), which says that the premisses must be *known* to be true; but, besides being a strengthening, this was also a change of emphasis, from theory to practice. In practice we often proceed on less than knowledge; namely, on more or less strong belief or acceptance. An argument that proceeds from *accepted* premises on the basis of an *accepted* inference-process may or may not be a good one in the full, alethic sense, but it is certainly a good one in some other sense which is much more germane to the practical application of logical principles.

And here it may be that the puristic logician will feel that I am lowering my sights, and declaring a preference for, or satisfaction with, arguments that persuade, as distinct from possibly unpersuasive arguments that are valid. This is a half-truth, and we must distinguish the different possible purposes a practical argument may have. Let us suppose, first, that A wishes to convince B of T, and discovers that B already accepts S: A can argue 'S, therefore T' independently of whether he himself accepts S or T and independently of whether S and T are really true. *Judged by B's standards*, this is a good argument and, if A is arguing with B and has any notion at all of winning, he will *have* to start from something B will accept. The same point applies to the inference-procedure. One of the purposes of argument, whether we like it or not, is to convince, and our criteria would be less than adequate if they had nothing to say about how well an argument may meet this purpose.

Conviction, of course, may be secured by threat, water-torture or hypnotism instead of by argument, and it is possible that Logic should have nothing to say about these means; but we can hardly claim that an argument is not an argument because it proceeds *ex concesso*, or that such arguments have no rational criteria of worth. We are, in fact, talking about the class of arguments that Aristotle called *dialectical*, and that Locke called *ad hominem*. The dialectical merits of an argument are, no doubt, sometimes at variance with its merits judged alethically or otherwise; but we would still do well to set down a set of criteria for them.

However, there is also more to be said against the alethic criteria and in favour of a set based on acceptability or acceptance rather than truth. The case in which Smith tries to convince Jones on grounds which Jones will accept but Smith may not, is, after all, somewhat less general than will satisfy us: we should consider, also, the case in which someone, *with good reason*, accepts a given set of premises and a given inference-process, and becomes convinced of a consequent conclusion. In other words, we should consider a case in which we are not at all tempted to make quasi-moral judgements. The question of whether there are any circumstances in which it is permissible to argue a case on someone else's grounds – though it would almost certainly be answered in the affirmative – is not really relevant, and we can dodge it by remarking on the relativity of the word 'accepted' which like 'known', is really doing duty for a range of concepts, 'accepted by me', 'accepted by you', 'accepted by modern Science' and so on. What good reasons various people may have for accepting various statements and procedures are, no doubt, themselves sometimes relevant to the worth of argument erected on them; but, if we are to draw the line anywhere, acceptance by the person the argument is aimed at – the person for whom the argument is an argument – is the appropriate basis of a set of criteria.

Somewhat more tentatively, one might push the claims of this reformulation even further. So long as it is the logic of practice that is being discussed, it is important to relate the concepts of truth, validity, and knowledge to dialectical concepts in the right way. Dialectical concepts have a certain claim to be considered as the fundamental ones, in that the 'raw' facts of the dialectical situation are that the various participants put forward and receive various statements. In the limiting case in which one person constructs an argument for his own edification – though we might follow Wittgenstein in finding something peculiar about this case[1] – his own acceptance of premises and inference are all that can matter *to him*; and to apply alethic criteria to the argument is surreptitiously to bring in the question of *our own* acceptance of it. When there are two or more parties to be considered, an argument may be acceptable in different degrees to different ones or groups, and a dialectical appraisal can be conducted on a different basis according to which party or group one has in mind; but again, if we try to step outside and adjudicate, we have no basis other than *our own* on which to do so. Truth and validity are onlookers' concepts and presuppose a God's-eye-view of the arena. When Smith and Jones argue and I am looking on, I can say to you 'Smith's argument is valid', or 'Jones's premises are

[1] I refer to the well-known 'private languages' argument, in *Philosophical Investigations*, 258, which can be adapted here.

false', in judgement of what I observe, and these statements are different from and irrelevant to anything about what Smith or Jones *accept*. But if *Smith* says 'S is true', the words 'is true' are empty and he might as well have said simply 'S'; and if he says 'The argument "S, therefore T"' is a good one' he might just as well have argued 'S, therefore T'. Used by *participants* in the argument, these terms cannot have the same function as for onlookers. And alternatively, if I as a former onlooker decide to intervene to give Smith the glad tidings that his argument is valid or Jones the news that his premisses are false, I am likely to find that I have become simply another participant in an enlarged dialectical situation and that the words 'true' and 'valid' have become, for me too, empty stylistic excrescences. To another onlooker, my statement that so-and-so is true is simply a statement of what I accept.

This point is of such fundamental philosophical importance that more needs to be said about it. The empty or, at best, parenthetical character of 'is true' and 'is valid' when applied to *my own* statements or arguments is paralleled by the similarly empty or parenthetical character of 'I think', 'I believe', or 'I accept'. Broadly, it would seem that the man who says 'S is true' or 'I accept S' might as well say simply 'S'; and, if he needs to paint in some subtle shade of meaning, can do so as easily by a gesture or an intonation as with extra words. What function do these extra words have? One answer is that one or other formula is essential to the case in which S is specified by description rather than explicitly: thus, I can affirm 'Jones's last statement but one *is true*' where it would not make sense to utter the verbless sentence 'Jones's last statement but one'. However, in the context of a discourse in which all parties understood the reference of the descriptive phrase, I could still dispense with 'is true' by substituting Jones's actual statement; and, in any case, we still have to discriminate between 'is true' and 'I accept'. Although my saying that X accepts S is not at all the same as my saying that S is true, my saying that *I* accept S seems, on the face of it, to have precisely the same function and practical effect.

The two formulations differ, as it happens, in a dialectical subtlety that involves not so much the speaker as the addressee. If Smith says 'S is true' and Jones agrees, he indicates that he,

Jones, accepts S in the same way as Smith does; but if Smith says 'I accept S' and Jones agrees straightforwardly, he indicates not that he accepts S but only that he understands that Smith does. Hence it makes a difference to the addressee, Jones, which form is used, and either form to some extent restricts the degrees of freedom of his reply. Knowing this, Smith himself will choose to say 'S is true' if he seeks acceptance of S by Jones, and 'I accept S' if he does not seek or expect this acceptance. Generally, a formulation in terms of what I accept, rather than in terms of what is true, does not issue so strong a challenge to the hearer.

So much for the participant; but consider, now, the position of the onlooker and, particularly, that of the logician, who is interested in analysing and, perhaps, passing judgement on what transpires. If he says 'Smith's premises are true' or 'Jones's argument is invalid' he is taking sides in the dialogue exactly as if he were a participant in it; but, unless he is in fact engaged in a second-order dialogue with other onlookers, his formulation says no more nor less than the formulation 'I accept Smith's premises' or 'I disapprove of Jones's argument'. Logicians are, of course, allowed to express their sentiments but there is something repugnant about the idea that Logic *is* a vehicle for the expression of the logician's own judgements of acceptance and rejection of statements and arguments. The logician does not stand above and outside practical argumentation or, necessarily, pass judgement on it. He is not a judge or a court of appeal, and there is no such judge or court: he is, at best, a trained advocate. It follows that it is not the logician's particular job to declare the truth of any statement, *or the validity of any argument*.

While we are using legal metaphor it might be worth while drawing an analogy from legal precedent. If a complaint is made by a member of some civil association such as a club or a public company, that the officials or management have failed to observe some of the association's rules or some part of its constitution, the courts will, in general, refuse to handle it. In effect the plaintiff will be told: 'Take your complaint back to the association itself. You have all the powers you need to call public meetings, move rescission motions, vote the manager out of office. We shall intervene on your behalf only if there is an offence such as fraud.'[1]

---

[1] See, for example, Head, *Meetings*, p. 101.

The logician's attitude to actual arguments should be something like this.

My statement of this position is, perhaps, more forthright than the support I can give it warrants; but some of those who disagree will still follow along with the idea of a weakening of the criteria of worth of an argument. The modified criteria, which I shall call *dialectical* ones, are formulated without the use of the words 'true' and 'valid'; or of the word 'known', which would imply truth. With this difference they run closely parallel to the epistemic criteria.

(D1) *The premises must be accepted.*

For 'accepted' one may read 'accepted by X', where the name of any person or group of persons may be put for 'X', provided the same substitution is made all the way through.

(D2, 3) *The passage from premises to conclusion must be of an accepted kind.*

This can be construed as implying reasonable immediacy.

(D4) *Unstated premises must be of a kind that are accepted as omissible.*
(D5) *The conclusion must be such that, in the absence of the argument, it would not be accepted.*

If we are prepared to countenance *degrees* of acceptance we can weaken this to: *The conclusion must be such that, in the absence of the argument, it would be less accepted than in its presence.* However, this makes the whole question of the worth of an argument a relative matter.

Now, an onlooker who wishes to apply these criteria to the assessment of an argument must decide from whose point of view he wishes it assessed – the arguer's, the addressee's, or his own. When an onlooker pretends to give an 'absolute' or 'impersonal' assessment the point of view is largely his own.

It is not uncommon for an argument to be assessed from a mixed point of view, by the construction of a hypothetical argument-situation having only some of the features of the actual one. Thus a logically-minded onlooker who judges 'That argument is valid' will frequently mean 'If I accepted those premises and did not accept that conclusion, that argument would persuade me'; or, given an example of an argument insufficiently spelt out, he will puzzle through it and conclude 'The argument is really valid', meaning, in our analysis, 'The argument is acceptable to me as supplemented with these steps'. We cannot, perhaps, legislate for the various special kinds of hypothetical argument-situation that a theorist can construct for himself, and we must content ourselves with regarding them as non-primary.

Why do I use the word 'accepted' in my primary formulation, rather than the word 'believed'? It would be natural to weaken 'S is known' to 'S is believed' rather than 'S is accepted'. My reason for preferring 'accepted' is that 'believed' is too much a psychological word, conjuring up pictures of mental states. I can *accept* something simply by putting on the appropriate linguistic performance; and this behavioural manifestation is the only necessary constituent of the argument-situation. I can conceive that a machine might be made to accept or reject arguments, though I would hesitate to describe it as having beliefs.

Now let us return to the three problems with which we started. I shall take them in reverse order, dealing, first with the thesis of Sextus and Mill that every argument is question-begging.

Sextus was interested in this thesis to support the sceptical view that no knowledge is possible: Mill, to destroy deduction as a source of knowledge by comparison with observation and analogy. In either case the thesis is an epistemological one, and, in retrospect, it can be seen that we would be right to expect a purely alethic analysis of argument to lack the richness to deal with it. Sextus and Mill could be regarded, in other words, as criticizing the alethic conception of argument in favour of one incorporating epistemological considerations. Aristotle, as we saw earlier in discussing his treatment of Begging the Question, was inclined to object to any argument that did not fit into the pattern he thought appropriate for the orderly or 'scientific' deduction of knowledge from first principles. Mill produces an empiricist's version of the same predilection; and Sextus the sceptic's version which disallows all knowledge and hence all argument.

The revised set of criteria of argument – whether epistemic criteria or dialectical – goes some distance towards meeting these three authors' objections to the alethic conception of argument, but then leaves them to go their separate ways; though with weakened separate theses. A question-begging argument has frequently been defined as one whose premisses are at least as much in need of proof as its conclusion, and this is precisely the kind of argument that is ruled deficient by criteria (E5) and (D5), which have no correlates in the alethic set. So long as we are using one of these more sophisticated sets of criteria, the truth is not that all good arguments are question-begging, but that none are. Moreover, not all alethically good arguments are question-begging; they are question-begging only if they fail to satisfy these additional non-alethic criteria. Put generally, the thesis of Sextus and Mill is hardly plausible once we have moved away from an alethic conception of argument.

Mill, of course, has done us a service in pointing out that there is a restricted conception of knowledge – the naïve empiricist conception – that gives approximately the result he states. If we are prepared to accept, first, that the only true knowledge is that which is obtained from direct observation and, secondly, that direct observation gives us always singular facts and never general, we shall have to regard any argument from a general premiss to a singular conclusion as wrong-headed and unscientific; and if we think thirdly, as Mill apparently does, that all deductive arguments are in the form of traditional syllogisms and, fourthly (and unhistorically), that syllogisms are inferences from general propositions to singulars, it will follow that anyone who increases his knowledge by the use of deductive argument is increasing it improperly or by the wrong means. This is not the place to debate the assumptions, but the special nature of them needs emphasis.

Different special stipulations are made by Aristotle and by Sextus. Mill, of course, has done little more than turn Aristotle on his head. For Aristotle, self-evident first principles are at the start of the epistemological argument-chain and all scientific knowledge is obtained deductively, though not all deduction is scientific. For Sextus the acquisition of true knowledge is not possible at all, and it would follow that not only deductive arguments are question-begging but every other kind too.

On the other hand, we need only relax these writers' special stipulations concerning the acquisition of knowledge to destroy their thesis. Let them say what they will about *true* knowledge; but, when it is knowledge of a more mundane variety that is being considered – and, particularly, if it is not so much 'knowledge' in any strict sense but beliefs, hypotheses, and theses – any kind of argument can be question-begging but no kind is more clearly so than any other.

Science and empirical (or other) method doesn't superannuate or by-pass dialectical criteria of argument. Even Science must progress by building on 'accepted' knowledge, and every scientific thesis needs to be supported by a dialectically sound case. It is perhaps even a danger for Science that it should be regarded as an enterprise built co-operatively on universally public empirical facts, rather than as a give-and-take market-place activity.

While we are berating philosophers for neglect of non-alethic criteria of argument we should take time off to accord special dispraise to the modern formal logician, whose field of view is perhaps more blinkered in this respect than that of almost any of his predecessors. His conception of argument is well illustrated by the formal concept of *proof*. A proof, for the formal logician, is a display of formulae of his system, either in a list or in a two-dimensional table, that satisfies certain structural properties, namely, that formulae later in the list or lower in the table are related by 'rules of inference' to relevant earlier or higher ones. Formal proofs, so conceived, have the virtue of precision but it is totally misleading to take such a proof as a model of rational argument.

In a formal proof the conclusion, or last formula, may be proved either absolutely, or relatively to certain other formulae higher in the table. If it is proved absolutely, this is either because no relevant formulae have been introduced unproved, or because those that have been introduced are axioms or (in some systems) because all unproved formulae have been prefixed in a 'conditionalization' process. If it is proved relatively to other formulae, then it would be said that the 'conditional' proof has been absolutely validated, which amounts to very much the same thing as saying that a certain formula of conditional form has been proved absolutely. The rules of inference and, if any, the axioms are supposed to be beyond question.

A proof, I take it, is just a knock-down argument; but this model of proof, far from setting a *high* standard of argument-worth for us, completely lets slip certain important desiderata. For example, it quite fails to ban circular reasoning for us, and one is encouraged to imagine that there is 'really nothing wrong' with using a formula to prove itself, or an axiom to prove an axiom, or a rule to prove a formula (such as *modus ponens*) interpretable as the expression of the rule. Equivocation is apparently also regarded as impossible, or the invalid arguments that it may lead to as 'formally valid'. The shortness of the steps and the transparency of the axioms and rules, whose rationale is the provision of a guarantee against error, is not only not a protection against these other sources of invalidity but a smoke-screen that can help them to slip through unnoticed; and it is not uncommon for the fussiness of a formal proof to defeat its own end by making it extremely laborious to follow, if not actually obscure. Yet in spite of it all it is a commonplace of modern Logic that highly paradoxical theorems have been 'proved' from harmless-looking axioms and rules. The complicated shuffle involving the construction of 'alternative' systems disguises the fact that nothing is proved absolutely at all, and that an unpalatable theorem can sometimes be a ground for going back and altering the axioms or rules.

The worst feature of this model is the appearance of definitiveness given to the concept of proof, and the impression that it is unrelated to the problem of filling out our knowledge or of actually convincing real people. Short steps are neither necessary nor sufficient to carry conviction: there is no guarantee of freedom from equivocation: conviction on one occasion or of one person is no guarantee of conviction on another occasion or of another: deductive 'proof' does not put a conclusion beyond rejection on other grounds. Formal validity, that is, is no guarantee of validity or vice versa.

This brings us to the second of our problems, that of 'non-deductive' argument-forms. Actually, it should be clear by now that we cannot always classify an argument uniquely as 'deductive' or 'non-deductive'. Arguments are more than mere inference-steps, and may have a structure with different elements in them. Nevertheless there are clear cases of arguments that are non-deductive: inductive arguments, statistical or probabilistic arguments, arguments from authority and arguments which rely on one or another kind of emotive appeal. Our earlier question, 'Are they really arguments?' can now be seen to be misplaced: certainly they are arguments, and if we are questioning any credentials at all we should rather question those of what often pass for *deductive* arguments.

When we have filled out the concept of an argument and realize that it is more than an inference schema, what, precisely, is the difference between deductive arguments and others? Does deductive argument give greater certainty? Not always: sometimes we rely for preference on the others. We rely on authority rather than deduction if we back the word of a distinguished mathematician against our own calculations; and on induction against deduction if we back a highly-confirmed experimental generalization against a theoretical prediction. (There is more, of course, to be said here, but it does not destroy the main point.) Is deductive argument more systematic than other kinds? Unfortunately, yes, for it is a scandal of modern Philosophy that, despite all the miles of writing, the question of induction or confirmation of hypotheses remains in such a state, utterly unmechanizable. But much has been done, and we should not regard deductive Logic as perfect. Is deductive validity, when we are sure we have the details correct, 100 per cent in a way other kinds of argument-worth are not? All sorts of things can go wrong with a deductive argument. Many a mathematician or logician has solemnly proved a contradiction, sure that he had the details correct. When we invoke the 100 per cent character of deduction as characteristic of it we are talking about pure theory, not practice.

Excessive obscuratism on this point is out of place: one is tempted to overstate a case purely because it is usually not stated at all. I shall assume that the reader really does know the difference between deductive arguments and others, and that I do not need to go into it. What is, above all, necessary is to dethrone deduction from its supposed pre-eminent position as a provider of certainty. This is not-at-all for Sextus's or Mill's reasons, but simply because we sometimes cheerfully and properly prefer other arguments against it.

The stumbling-block for many people is the mistaken idea that a good deductive argument *compels* the acceptance of a conclusion which, in turn, entails unequivocal action on it. Nothing has been said in this chapter about the rationality or otherwise of accepting the conclusions of good arguments and of acting on them, and it should be clear that this is not an entirely simple matter – when, for example, there are good arguments pointing in opposite directions. Once this question is separated, it should be clear that the very existence of different argument forms is a part of the problem with which, in the long run, the logician must deal, since there must be rules for weighing one against the other. Here, it must be sufficient to protest against the theory that the weighting of deductive arguments is determinate, and infinite, and that of inductive arguments totally the reverse.

Now let us turn to the problem of 'nailing' a fallacy. Is it a genuine problem? Good arguments (we hear the plaintiff saying) should be seen as good by all reasonable people, whereas some people refuse to be impressed and our knock-down arguments leave them still standing up. What is a rational man to do about those who are irrational and will not admit it?

This complaint must be dismissed as frivolous. It amounts to the demand that there should be a precise equation between *logical soundness* and *practical efficacy*: Right must be Might. And the answer to this demand is, first, that there is no royal road to success in practical dialectics; but, secondly, and most importantly, that no argument, even when wilful sophistry is set aside, ever *settles* a dispute once and for all, beyond the possibility of being reopened.

What argument ever is knock-down? Some, of course, are sometimes accepted as being so. But it is not at all unusual to find that an apparently knock-down argument – which, perhaps, satisfies all of somebody's rules of validity – is later found to be faulty. Either it is discovered that one of the premises was untrue or insufficiently substantiated, or it is found that there was an equivocation on some term, or that the question was begged, or that there was a confusion concerning exactly what was being proved; or, though perfectly valid and drawn from true premises it was not straightforwardly drawn and should have had some filling-out or marginal explanations; or, though it was valid and from true premises, the arguer's or hearer's reasons for thinking it so were misplaced, the actual truth or validity being only accidentally achieved. Or it is discovered that there are other powerful arguments contradicting the conclusion reached and that a reappraisal of the earlier argument should be undertaken in spite of its strength; or that there is an unexpected repugnance between the conclusion and newly-discovered other facts; and so on, virtually *ad infinitum*.

This sceptical doctrine needs to be balanced, of course, with the well-known countermoves to scepticism. Do I really think, in the case of such-and-such well-accepted arguments, that there is any likelihood of reversal? When someone describes an argument as 'knock-down' and it seems to him, to me and to everyone else to be so, are we wrong so to describe it? No. The use of the term remains what it was. But if the philosophical point has been well made, something follows about the attitudes that should be taken towards the concept of argument, and those that should not be. Many of the latter are current.

CHAPTER 8

Formal Dialectic

Let us explore the second half of the definition of fallacy a little a little further and be clearer what it is for an argument to *seem valid*. The term 'seem' looks like a psychological one, and has often been passed over by logicians, confirmed in the belief that the study of fallacies does not concern them. The arguments that tell against a psychological interpretation of logical terms. however, tell also against this supposition. That B seems *to me* to follow from A, when in fact it does not do so, implies that I am guilty of error, but it is not in itself grounds for calling the argument from A to B a fallacy. John Smith may believe that it follows from the state of the mining market that the moon is made of green cheese, and, if he so argues, his argument is very likely invalid; but if we discover that his belief is an isolated one without a rationale we shall be inclined to withhold the description 'fallacy' and say no more than that it has no logical foundation. Some similar reaction would be appropriate even if we were to find that a large number of other people shared his belief. We might, it is true, get round to speaking of the belief as a 'popular fallacy', in the sense in which we regard it as a onetime popular fallacy that the earth was flat, but we have put this sense of the word 'fallacy' behind us. An *unsystematic* belief is not a candidate for the title 'logical fallacy' even when it is in the form of an implication and widely held.

To justify the application of the tag 'fallacy', the seemingvalid must have a quasi-logical analysis. But what is the quasi-Logic within which this analysis is performed?

One kind of case in which we do not hesitate to speak of

fallacy is that in which we are confronted with a false logical doctrine. If the invalidity of Smith's argument were due to his thinking that universal affirmative propositions are convertible, or that it is permissible to permute mixed quantifiers, we should identify his fallacy in just these terms. We do not demand, of course, that the fallacy-monger be capable of a clear and explicit formulation of his false principle: it is sufficient that we ourselves discern the operation of such a principle in his reasoning. To weaken the requirement a little further: it is common enough that what we discern is neglect of a true principle rather than conformity to a false. The important thing is that the invalidity must be systematic, and its source re-identifiable in different instances. Once this is said, it is clear that it need not, and ultimately must not, be described psychologically. A *formal* fallacy is an invalid argument generated by just such a false logical doctrine, and so there is nothing psychological about the seeming-valid except the fact that, for practical utilitarian reasons, we tend to confine consideration to cases in which the false doctrine is one that is liable to be actually held or followed by real people.

There are two ways in which fallacies may fail to be 'formal': they may either, like Begging the Question, be such that they do not rest on formal invalidity; or they may consist of arguments that are, indeed, formally invalid but are such that there is no possible (spurious) formal principle which generates them and gives them their seeming-validity. How are we to analyse these?

The answer, in both cases, is that we need to extend the bounds of Formal Logic; to include features of dialectical contexts within which arguments are put forward. To begin with, there are criteria of validity of argument that are additional to formal ones: for example, those that serve to proscribe question-begging. To go on with, there are prevalent but false conceptions of the rules of dialogue, which are capable of making certain argumentative moves seem satisfactory and unobjectionable when, in fact, they conceal and facilitate dialectical malpractice. Most of Aristotle's Fallacies Outside Language, and many of the stragglers that have been imported into other lists of Fallacies, are analysable within Dialectic in a way they would not be in Formal Logic. The Fallacies Dependent on Language are in a slightly different category and we shall set them aside for separate consideration in the next chapter.

Formal Dialectic, it should be added, does not have as its sole justification the analysis of fallacies; least of all, of those of the accepted list. It is the discipline to which the discussions of fallacies in the textbooks obliquely introduce us, and which represents the unstated *raison d'être* of those discussions; and it will probably superannuate them as Formal Logic has superannuated Topics. Its relation to the study of fallacies is, admittedly, not quite straightforward, and we find elements relevant to it, from time to time, in positive theories of reasoning also; but, if the scattered survival of discussions of fallacies requires a justification other than in terms of entertainment value, this is what it must be. We need to see our reasoning in the kind of context within which, alone, these faults are possible.

Let us start, then, with the concept of a *dialectical system*. This is no more nor less than a regulated dialogue or family of dialogues. We suppose that we have a number of participants – in the simplest case, just two – to a debate, discussion or conversation and that they speak in turn in accordance with a set of rules or conventions. The rules may specify the form or content of what they say, relative to the context and to what has occurred previously in the dialogue. They govern the speaker's language and his logic, as well as a number of features of his discourse which are not normally studied under either of these headings.

In our present discussion we shall not be concerned to consider any contact of the dialogue with the empirical world outside the discussion-situation. It is true that the possibility of such contact is often germane to the formulation of dialectical rules and that there are some dialectical phenomena – ostensive definition, the language of perception, commands, and others – which cannot be profitably treated in its absence; but our present range of problems does not call for this generality. Here we merely note the omission; which otherwise might leave us open to the kind of criticism levelled against medieval Dialectic by the Renaissance. An interest in Dialectic has frequently been associated, in philosophical history, with a claim to discover knowledge by the use of purely dialectical methods, but this is no part of our present plan.

The study of dialectical systems can be pursued *descriptively*, or *formally*. In the first case, we should look at the rules and conventions that operate in actual discussions: parliamentary debates, juridical examination and cross-examination, stylized communication systems and other kinds of identifiable special context, besides the world of linguistic interchange at large. A formal approach, on the other hand, consists in the setting up of simple systems of precise but not necessarily realistic rules, and the plotting of the properties of the dialogues that might be played out in accordance with them. Neither approach is of any importance on its own; for descriptions of actual cases must aim to bring out formalizable features, and formal systems must aim to throw light on actual, describable phenomena. As a matter of emphasis, however, I shall lean towards a formal approach in what follows, since the practical material we aim to illuminate – fallacious argumentation – has already been sufficiently described.

Dialectic, whether descriptive or formal, is a more general study than Logic; in the sense that Logic can be conceived as a set of dialectical conventions. It is an ideal of certain kinds of discussion that the rules of Logic should be observed by all participants, and that certain logical goals should be part of the general goal.

The concept of a dialectical system is, at first, quite general and there are many systems that are of no interest whatever to the logician. For example, we can imagine a dialogue consisting of interchange of statements about the weather. Even remark-trading of this kind, however, is not entirely without interest if certain additional requirements are imposed: that a speaker's remarks should be consistent one with another, and that they should be without repetition or, perhaps, mutual implicative relationship; and that there might be some specification of an interaction between the speakers. We might imagine that, in certain circumstances, a speaker is obliged to indicate agreement or disagreement with a preceding remark of the other speaker, as if it were also a question. In fact a question-and-answer system, in which A asks questions and B must provide syntactically correct answers to them, is really simpler than this, since a questioner is not directly involved in the matter of consistency; but, even so, provides a generic dialectical setting for the decision-problems of all the formal logical calculi. A speaker who is obliged to maintain consistency needs to keep a store of statements representing his previous commitments, and require of each new statement he makes that it may be added without inconsistency to this store. The store represents a kind of *persona* of beliefs: it need not correspond with his real beliefs, but it will operate, in general, approximately as if it did. We shall find that we need to make frequent reference to the existence, or possibility, of stores of this kind. We shall call them *commitment-stores*: they keep a running tally of a person's commitments.

Rules may prescribe, prohibit, or permit; may be directed to particular people, who play roles in a dialogue; and may be conditional on any feature of the previous history of the dialogue. We shall normally avoid 'permissive' rules, adopting the liberalistic convention that anything is permitted that is not specifically prohibited. (It would, of course, be dualistically possible to frame all rules as permissive ones and prohibit everything they did not license.) The things prescribed or prohibited are linguistic *acts* of the person concerned, perhaps including the null-act; and a linguistic act is defined as the speaking of a locution in a given language. Since we are concerned mainly with two-person dialogues we can dispense with the phenomenon of discriminatory direction of locutions to one person rather than another, and assume that all locutions are directed to the other participant. The locutions we are concerned with generally represent *statements* or *questions* drawn from some prescribed range, but there will be certain others of a procedural nature; and, in any case, it should be borne in mind that it is ultimately their role in the dialectical system that gives sentences this kind of character, rather than the other way about.

We shall assume that speakers take turns politely, but this does not rule out the possibility that a given contribution by a given speaker may be analysable into two or several individual sentences. If it were necessary to be precise on this point we could build into each system a set of rules designed to determine who speaks when: for example, we could stipulate that each contribution of each speaker, except the final contribution of the dialogue, should end with the special locution 'Over!', and that each contribution except the first should immediately follow the saying of 'Over!' by the previous speaker. This would not give protection against filibusters but other means could be devised. We shall not need to involve ourselves in any of these matters in this book.

Rules, then, are of the general form 'If C is the case, sentences of the set S are prohibited for person P'; where prescribing A is the same thing as prohibiting everything else but A. Here C is a specification of a feature of the previous history of the dialogue. More adequately we might define *incomplete dialogue* as, say, any dialogue not terminating in some standard way, such as with the word 'Finish'; and then let C represent the prior occurrence of some one of a specified set of incomplete dialogues with P as a participant. In practice we are generally interested in rather easily specified sets, as in 'If the contribution of the other party was a question of the form S?, P must now say S or *Not-S* or *I don't know*'. Very frequently, too, the past history of the dialogue is sufficiently summed up by the traces it has left in the contents of commitment-stores.

In general, we need to specify two *languages*: the *object-language*, which is the one used by the speakers in the dialogue, and the *rule-language*, which is the language used in stating the rules and contains means for describing features of dialogues in conditionalizing these rules. Again, for present purposes, we shall not need to be very strict about either of these language-specifications; but both will, in some circumstances, be of relevance to the properties of systems.

A system is *rule-consistent* if its rules are such that it can never arise that one and the same act is both prohibited and prescribed; or, equivalently, there is no circumstance in which all possible acts (including the null-act) would be prohibited. Rule-inconsistency does not always matter in a practical system since sometimes the person who stands to be inconvenienced can see the contingency coming and take steps to avoid it. However, it is very easily avoided in such cases by minor reformulation.

Rule-consistency is a concept applicable to all systems. Questions of consistency may also enter at the level of the object-language, provided that, as is usually the case, we envisage this as containing *statements*, in some reasonably normal sense of that word. We shall say that a system is *semantically consistent* if it is never unconditionally possible for a speaker, in a single locution, or in several separate locutions taken together, to be forced to utter a contradiction; and *semantically unforced* if there is no non-tautological statement that a speaker can be unconditionally forced to utter, or any set of statements of which he can be forced to utter one, other than a set whose disjunction is a tautology. It is *semantically unforced with respect to a given (complete or partial) evaluation* – namely, with respect to some allocation of truth-values to contingent statements of the object-language – if no speaker can be unconditionally forced to utter a falsehood. The concept of a set of rules relative to an evaluation could be an important one in some connections, where the evaluation may be taken as representing shared sensory information and beliefs. A system is *semantically open* if there is no statement at all, even a tautological one, which a speaker can unconditionally be forced to utter, nor any set of statements of which he can be unconditionally forced to utter one. A semantically open system is semantically unforced. A semantically unforced system is one that is unforced relatively to every evaluation. Semantically unforced systems are semantically consistent. It should be re-emphasized that these characterizations are restricted to systems in which *statements* occur, and that this stipulation, in a sense, puts the cart before the horse. In the long run, whether a given locution is or is not a statement, question or the like depends upon its place in a dialectical system, and not vice versa. A *statement* is, ultimately, a locution which has permissibility-rules relating it to an effective commitment-store and governs the permissibility, in similar ways, of subsequent locutions; a *question* is a locution with respect to which there is a rule requiring one of a specified set of statements as subsequent locution of the other speaker, and so on.

It frequently happens that there are rules at different levels: at one level there are rules which specify, syntactically, what does or does not represent a dialogue of the system in question, whereas, at another level, there are rules which distinguish some dialogues among them as being more rational, or 'better', or a 'win' for this speaker or that, or as otherwise belonging to some more restricted class than dialogues as a whole. Moreover, it is not always clear how we should apply these distinctions. Take the matter of consistency: we have a choice whether to regard an inconsistent statement as unsyntactical, and hence as impossible within a given system properly defined, or to regard it as a perfectly possible locution which is simply of a kind towards which we take a particular attitude. It is perhaps less obvious that we have, in the long run, the same option with regard to various ranges of clearly unsyntactical, ill-formed and meaningless utterances. In ordinary speech, if a participant at some moment gives vent to an unidentifiable noise, we may ignore it and treat it as being no part of our conversation; but we may, depending on details difficult to regulate, choose to regard it as a locution to which we make some response such as 'What do you mean by that?' or 'That is not an answer to my question'. It is a weakness of formal work in these fields that it seeks to draw a hard line where none exists; but again, having noticed this, let us pass on.

We are ready to consider an example; and the one that recommends itself is the Obligation game. This is less fruitful than some other possibles such as the Lincoln's Inn or Buddhist varieties in throwing light on Fallacies, but it is the simplest of the various systems that might interest us. I shall aim to give a version which captures what I take to be the essence of the game in its simplest form, without claim to historical accuracy in detail.

The Obligation game is played by two people, called the Opponent and the Respondent. The language used is a finite propositional language based, say, on elementary propositions $a_1, a_2, \ldots a_k$ and truth-functional operators, supplemented with several special locutions. (In place of propositional calculus we could substitute any other finite language of sufficiently normal type; for example, lower predicate calculus on a universe with finitely many individuals and limited variety.) We shall not formalize the rule-language. The Opponent speaks first and his first locution has three parts:

(*a*) The words 'Actual fact:', followed by an evaluation on the language, consisting, say, of a state-description $b_1.b_2 \ldots b_k$ where each b_j is a_j or $-a_j$. We shall call this statement '*B*'.

(*b*) The word 'Positum' followed by a contingent statement inconsistent with *B*. We shall refer to this statement as '*C*'.

(*c*) The words 'Propositum 1' followed by a statement, to be referred to as 'P_1'.

The Respondent's first locution R_1 consists of either P_1 or its negation $-P_1$; and, in general, each of the Respondent's contributions R_n ($1 < n < m-1$) consists of repeating the preceding locution of the Opponent, or stating its negation. The Respondent's locutions may be reckoned 'correct' or 'incorrect' in accordance with a subsidiary rule to be detailed in a moment. The Opponent's contributions O_n ($2 < n < m-1$) are of the form: the words 'Propositum *n*' (for whatever *n*) followed by a contingent statement P_n. The Opponent's final locution O_m is the words 'Win and Finish' if R_{m-1} was incorrect, 'Resign and Finish' if R_1 is correct and $m=11$, i.e. if the Respondent has survived ten *proposita*.

The answer-rule is as follows. Associated with the game is a commitment-store for the Respondent, consisting of the conjunction of the *positum* and all the Respondent's answers to date: that is, after the Respondent's *n*th locution the store contains C_n, where $C_{j+1} = C_j.R_j$ for each $j = 0, 1, \ldots m+1$. The Respondent's locution R_n is *correct* if it is either (i) implied by C_{n-1}, or (ii) consistent with C_{n-1} and implied by *B*; otherwise it is *incorrect*.

It can easily be checked that the system is rule-consistent: the only point of any doubt concerns the Opponent's final move. In assessing semantic properties we would do well to regard the Opponent's *proposita* as yes–no questions rather than as statements, on the grounds that they raise no question of commitment or consistency: only the Respondent's 'answers' do this. The system is then semantically consistent, open and unforced.

A slightly different system, which results from writing into the rules the stipulation that the Respondent must always give the 'correct' answer, can also be proved rule-consistent, as a simple consequence of the semantic consistency of the system just outlined. (The option 'Win and Finish' on the part of the Opponent will, consequently, never occur.) This system is semantically consistent but it is neither open nor, in general, unforced. It is, however, unforced, after O_1 has occurred, with respect to a certain valuation constructible in terms of B and C_0 as follows: to each statement S of the language is allotted the value 'true' if it is implied by C_0, and 'false' if it is counterimplied by C_0: otherwise it has the value 'true' if it is implied by B, and 'false' if it is counterimplied by B. We might hence regard the system as semantically unforced with respect to an evaluation if we are prepared to regard the 'actual fact' and *positum* as characteristic of the game and hence not part of a forcing move on the part of the Opponent. That the modified system is semantically consistent was known to William of Sherwood (?), who regards the Obligation game as illustrating the principle 'From the possible nothing impossible follows'.

Now let us make another modification to the system, of a kind implicit in one of William's (?) examples. Let the Opponent's *proposita* $P_1, \ldots, P_n$ consist not of single statements but of non-empty sets of statements, and let each 'answer' of the Respondent consist *either* of the conjunction of all the statements in the proferred set, *or* of the conjunction of their negations. For example, let P_j be the set of statements $\{p_1, p_2, \ldots p_i\}$ and hence let R_j be either $p_1.p_2 \ldots\ldots p_i$ or $-p_1.-p_2 \ldots\ldots -p_i$. Let us consider in turn what happens if we do *not* write in the requirement of 'correctness' of the Respondent's locutions, and what happens if we do so. In the first case the system is still clearly rule-consistent, but it is not semantically consistent since the *propositum* $\{p, -p\}$, for any p, is such that neither the conjunction of the statements nor the conjunction of their negations is consistent; and even the stipulation that, say, the statements $p_1, p_2, \ldots p_i$ of any set P_j must be mutually indifferent would still leave the system a semantically forced one. In the second case – if, that is, the 'correctness' requirement is written in – the system is not rule-consistent: for it is impossible to achieve a 'correct' R_j in answer to a P_j of the form $\{p, -p\}$; and even if statements of a set are mutually indifferent it is impossible to achieve a 'correct' response to *both* of $\{p, q\}$ and $\{p, -q\}$.

The vagaries of this system are a result of the demand that the Respondent give the same evaluation to each of a set of not-necessarily-equivalued statements. They therefore illustrate the operation of the Fallacy of Many Questions.

A similar result would be achieved by the presentation of *proposita* as 'biased questions' in the following form: each P_j is a set of two or more (contingent) statements and each R_j is a selection of *one* of the statements from the set. Such a system is, in any case, a systematically forced one, since any statement p may be foisted on the Respondent by giving him, first, the choice $\{p.q, p.-q\}$; and, if 'correctness' is written in, it is rule-inconsistent since a subsequent 'choice' may be $\{-p.r, -p.-r\}$. A *propositum* of the form $\{p.q, -p.-q\}$ in this system is equivalent in operation to one of the form $\{p, q\}$ in the previous one.

It follows that a 'biased question' is equivalent to a 'multiple question'.

Now, preparatory to considering some other systems, let us describe in more detail the organization and operation of commitment-stores. We have said that these stores contain *statements*, but it is not quite clear what this means, and it is better to be quite explicit and say that they contain *statement-tokens*, in the language of the dialogue or an equivalent one. We can conceive a commitment-store physically as a sheet of paper with writing on it, or as a section of the store of a computer. As a dialogue proceeds items are periodically added and, in some circumstances, deleted. A statement S is added, that is, whenever the speaker 'makes' it, or otherwise incurs it as a commitment such as by having it made to him and taking no steps to deny it; and it is deleted, perhaps with consequential alterations, if he 'retracts' it.

The sum total of the statements in the store at any time is the speaker's *indicative commitment*. (I say 'indicative' because, in other contexts which will not concern us here, it would be necessary to consider also other kinds of commitment such as imperative and emotive.) For some purposes we can consider the various statement-tokens as effectively conjoined into a single large one, but for other purposes it is necessary to think of them as distinct.

At first sight we would suppose it to be a requirement of the statements in a commitment-store that they be *consistent*; but, on reflection, we may come to think that, although there does exist an ideal concept of a 'rational man' which implies perpetual consistency, the supposition is by no means necessary to the operation of a satisfactory dialectical system. In fact, even where our ideals of rationality are concerned, we frequently settle for much less than this: a man is 'rational', in a satisfactory sense, if he is capable of appreciating and remedying inconsistencies when they are pointed out. We should reflect, too, that consistency presupposes the ability to detect even very remote consequences of what is stored, and that this would itself make nonsense of certain kinds of possible dialectical application. Could we model a discussion, between mathematicians, of the validity of a certain theorem, if we had to model the mathematicians themselves as all-seeing? In a discussion of a proof a participant may be committed to one step, but not yet committed to the next, which may be still under discussion. This, at least, is so in the sense of 'commitment' relevant to dialectical systems: others may use what sense they may.

At the same time, it is clear that certain very immediate consequences of S may be regarded as commitments if S is a commitment, and that flat and immediate contradiction between, say, S and $-S$ is not necessarily to be tolerated. There is a line to be drawn, and this will be a matter for regulation in a given system. A similar, logically rather difficult, question concerns retraction. When a participant 'changes his mind' about a statement S and either simply retracts it or replaces it by its negation it is simple enough to specify that S itself be deleted from his commitment-store; but in practical cases there will often need to be compensating adjustments elsewhere. What is to happen to a commitment S which *implies* T, when T is replaced by $-T$? If S is $p.q$ and T is p, it would seem that we should be left with $-p.q$; but $p.q$ is equivalent to $p.(p \equiv q)$ and the same reasoning would lead us here to $-p.(p \equiv q)$, which is equivalent to $-p.-q$. The answer is that the concept of 'retraction' is not as simple as it appears on the surface and, again, that rules need to be laid down in particular systems.

It may make these points a little easier to digest if we emphasize that a commitment is not necessarily a 'belief' of the participant who has it. We do not believe everything we say; but our saying it commits us whether we believe it or not. The purpose of postulating a commitment-store is not psychological. Although, presumably, the brain of an actual speaker must contain some remote analogy of a commitment-store, it contains much else besides; and the primary theoretical job that commitment-stores do for us is to provide us with a dialectical definition of *statements*.

In order to illustrate that the problems of the organization of commitment are not insoluble, let us set up a simple dialectical system in which one possible solution is provided. It will, incidentally, be a system in which the concept of Argument is realized, and within which we can model a number of Fallacies.

White and Black are two participants and the language is primarily that of the statement calculus, or some other suitable logical system with a finite set of atomic statements. The *axioms* of the language are contained in both participants' commitment-stores from the beginning, and are marked in some way to indicate that they occupy a privileged position. White moves first but the system is otherwise symmetrical between them. Locutions may consist of the following forms (I use capital letters S, T, U, ... as variable statement-names and other symbols and words autonomously):

(i) 'Statement S' or, in certain special cases, 'Statements S, T'.

(ii) 'No commitment S, T, ... X', for any number of statements S, T, ... X (one or more).

(iii) 'Question S, T, ..., X?', for any number of statements (one or more).

(iv) 'Why S?', for any statement S other than a substitution-instance of an axiom.

(v) 'Resolve S'.

This is a 'Why–Because system with questions'. A simple 'Why–Because' system (the simplest feasible) would omit locutions of type (iii). A simple 'question and answer' system could be built omitting locutions of type (iv).

Rules of the system fall into different categories. I shall first lay down a set of rules which do not depend on commitment-store operation and may be regarded as 'syntactical'; then describe the operation of the commitment-store; and later, in the subsequent discussion, suggest additional rules that are framed in terms of the speaker's and hearer's commitments.

Syntactical rules

S1. Each speaker contributes one locution at a time, except that a 'No commitment' locution may accompany a 'Why' one.

S2. 'Question S, T, ..., X ?' must be followed by
  (*a*) 'Statement —(SvTv...vX)'
  or (*b*) 'No commitment SvTv...vX'
  or (*c*) 'Statement S' or
    'Statement T' or
    ——— or
    'Statement X'
  or (*d*) 'No commitment S, T, ..., X"

S3. 'Why S ?' must be followed by
  (*a*) 'Statement —S'
  or (*b*) 'No commitment S'
  or (*c*) 'Statement T' where T is equivalent to S by primitive definition.
  or (*d*) 'Statements T, T ⊃ S' for any T.

S4. 'Statements S, T' may not be used except as in 3(*d*).

S5. 'Resolve S' must be followed by
  (*a*) 'No commitment S'
  or (*b*) 'No commitment —S'.

Commitment-store operation

C1. 'Statement S' places S in the speaker's commitment store except when it is already there, and in the hearer's commitment store *unless* his next locution states —S or indicates 'No commitment' to S (with or without other statements); or, if the hearer's next locution is 'Why S ?', insertion of S in the hearer's store is suspended but will take place as soon as the hearer explicitly or tacitly accepts the proferred reasons (see below).

C2. 'Statements S, T' places both S and T in the speaker's and hearer's commitment stores under the same conditions as in C1.

FORMAL DIALECTIC

C3. 'No commitment S, T, ..., X' deletes from the speaker's commitment store any of S, T, ..., X that are in it and are not axioms.

C4. 'Question S, T, ..., X?' places the statement SvTv...vX in the speaker's store unless it is already there, and in the hearer's store unless he replies with 'Statement —(SvTv...vX)' or 'No commitment SvTv...vX'.

C5. 'Why S?' places S in the *hearer's* store unless it is there already or he replies 'Statement —S' or 'No commitment S'.

The following is a specimen of a dialogue (with A, B, C as atomic statements and the two speakers distinguished by roman and italic type):

Question A, *—A?* Statement *A.* Question B, *—B?* Statement —B. Statement B[1]. *Statement —B*[2]. Why *—B?* *Statements A, A ⊃ —B.* No commitment A ⊃ —B; why A ⊃ —B?[3] *No commitment A ⊃ —B*[4]. Statement B[5]. *Why B?* Statements A, A ⊃ B. *Why A ⊃ B?* Statements —A, —A ⊃ (A ⊃ B). *No commitment —A ⊃ (A ⊃ B)*[6]. Resolve A. *No commitment —A.*[7]. Statement —A[8]. *Resolve A.* No commitment A[9]. *Why —A ⊃ (A ⊃ B)?* Statement —A ⊃ (—AvB) [10]. *Why —A ⊃ (—AvB)?* Statements —A ⊃ (Bv—A), (—A ⊃ (Bv—A)) ⊃ (—A ⊃ (—AvB))[11]. *Question A. C, —A.C?* No commitment A.C v —A.C[12]. *Statement C.* No commitment C. *Statement C...*[13]. No commitment C. *Why B?* Statements B, B ⊃ B. *No commitment B; why B?* Statements B, B ⊃ B...[14]. *Statement B*[15]. Resolve B[16]. *No commitment —B.*

Comments: (1) White (roman type) has accepted A but is not going to accept —B. (2) But Black (*italics*) must not accept B from White: he can reiterate —B or could equivalently say 'No commitment B'. (3) White has already accepted A and tacitly accepts it here again. But he *both* rejects *and* requires proof of A ⊃ —B. (4) Black retracts under pressure. (5) White doesn't need to reiterate B, but it does no harm. (6) Black has refused to accept —A ⊃ (A ⊃ B) in spite of the fact that it is a tautology, but has tacitly accepted White's self-contradiction —A. (7) White charges him with it and he retracts. (8) It is again unnecessary for White to reiterate, but he is aggressive in these matters. (9) Black says '*Tu quoque*' on the contradiction and White chooses the other leg. They are now both internally consistent but are at odds, though neither subsequently notices. (10) By definition of '⊃'. (11) The first of these is a substitution-instance of an axiom and cannot be further challenged. The second could be challenged but could be supported eventually by a valid proof. (12) The presupposition of the question is disallowed. (13) This can go on for ever. (14) So can this. (15) '*All right, have it your own way!*' (16) 'But earlier you said . . .'.

So far the system is clearly rule-consistent. It is also semantically open since it is possible for either speaker to say 'No commitment S', for some S or other, at any time, and hence to avoid all commitment. The inerasable axioms influence the dialogue in no way except in prohibiting 'Why S ?' when S is a substitution-instance of an axiom. The commitment-store rules are, of course, quite irrelevant to the conduct of the dialogue until we have laid down further rules framed in terms of commitment. At the same time the concept of commitment has a clear intuitive meaning and it is instructive to follow the state of the participants' respective commitments through the above sample dialogue. The solution of the problem of retraction, it will be noted, is to separate off the question of consistency, so that retracting S does not necessarily involve retracting anything related to S; that is, not even equivalents such as — —S and S.S. It would be possible to relax this stipulation to some extent but particular suggestions for so doing need to be looked at on their merits. Certain extensions of the commitment-mechanism in other directions will be touched on in a moment.

The syntactical rules for questions provide an escape for the addressee of a biased question: 'Question S, T, . . ., X ?' can be answered 'No commitment SvTv . . . vX', or 'Statement —(Sv Tv . . . vX)'. The commitment-rules provide that it *must* be so answered if the addressee is to escape commitment to SvTv . . . vX: thus 'No commitment S, T, . . . X' leaves him committed to the disjunction and merely escapes commitment to any particular one of the disjuncts. Is this reasonable, and is this all that needs to be done to model the phenomenon of the biased question? There is at least one further possibility. It might be felt that questions should not be used to make statements, and should not, therefore, themselves commit speaker or hearer in the way that is characteristic of statements. To achieve this, the system could be strengthened by the addition of a rule as follows:

'Question S, T, ..., X?' may occur only when SvTv ... vX is already a commitment of both speaker and bearer.

Such a rule could seriously impede the asking of questions since even 'Question A, —A?' would require the prior establishment of Av—A; but it is possible to envisage that, with other adjustments, the admittedly restrictive system that resulted might be seen as representing some kind of ideal of rationality in the use of biased questions. Weaker rules might be considered, such as that the disjunction of answers be a commitment of at least the addressee, or of at least the speaker. At all events rules of this kind are what are required to banish from the system various versions of the Fallacy of Many Questions.

These rules differ from the earlier ones in a way we might describe by saying that they have a discretionary character. What we called 'syntactical' rules define the scope of the system: those now being discussed seem aimed rather at improving the minimum quality of the resulting dialogues or degree of *rapport* between the participants. It is possible to use biased questions and not lead or be led astray by them; but these rules are what are required to provide a guarantee of unexceptionable use.

Similar rules framed in terms of commitments could be added to give a particular character to statement-making or question-asking. If we regard the sole function of statements to be the giving of information, we shall restrict their use with the rule:

'Statement S' may not occur when S is a commitment of the hearer.

Similarly, of course, for 'Statements S, T'. This is appropriate to a particular kind of statements which we might call 'notifications', and needs to be matched by a view of questions as 'inquiries', with the rule

'Question S, T, ..., X?' may not occur when any of S, T, ..., X is a commitment of the speaker.

On the other hand, statements have another function in language, namely, as 'admissions'; and questions a corresponding function as no-liability admission-elicitations; and in this case what is stated may or may not be a prior commitment of the hearer but is definitely not one of the speaker, whereas a question may or may not have an answer among commitments of the speaker but definitely does not have one among those of the hearer. Hence the rules given need to be weakened at least to

'Statement S' may not occur when S is already a commitment of both speaker and hearer.

and similarly for 'Statements S, T'; and

'Question S, T, ..., X?' may not occur when any one or more of S, T, ..., X, is a commitment of the speaker and any one or more a commitment of the hearer.

These rules are still strong enough to eliminate certain possible repetitious inanities such as 'Question A, B? *Statement A.* Question A, B? *Statement A* ...'.

These rules might be acceptable in a pure question-and-answer system but in the present system at least the first of them is still too strong, since it would prohibit the use of axioms in the justification of their consequences. Let S be an axiom and T a formula equivalent to it by definition: then we want to license 'Statement T. *Why T?* Statement S' but, since S is inerasably in both commitment-stores, 'Statement S' is permanently forbidden. The rule would need to be weakened even further at least to the extent of excluding statements made in answer to 'Why' locutions. The rule in respect of questions can probably be retained.

Another kind of repetitiousness, exemplified by 'Statement A. *No commitment A.* Statement A. *No commitment A*', is more difficult to ban. What is needed is an elaboration of the commitment-store operation to have it register second-order commitments of the form 'No commitment A', when a locution of this form is uttered. This is not an impossible extension but we need not pursue it here.

A possible condition on answers to questions concerns (what I shall call) 'whole truth commitment'. When it is said of an answer that although it is true it is not the 'whole truth', what must be understood is that the answer should properly be replaced by a stronger one which gives additional information; and that, by giving the weaker answer, the answerer has 'implied' that he is not in a position to give the stronger one or that the stronger one would be false. It would be possible to build 'whole truth commitment' into answers by treating every answer to any question as if it were accompanied by a 'No commitment' locution with respect to such other possible answers as are not actually implied by the one given. Certain restrictions on the formulation of questions are entailed by this stipulation, but the result would be at least a partial explication of what we mean by 'the whole truth'.

Can we introduce rules corresponding with the criteria of argument in the previous chapter? Argumentation appears in this system only in a rudimentary form, since the statement or statements which follow a 'Why' locution are not necessarily to be regarded as the ultimate premisses of an argument and the depth to which the participants probe is unregulated. If the utterer of a 'Why' locution is regarded as inviting the hearer to convince him, a reasonable rule is

'Why S?' may not be used unless S is a commitment of the hearer and not of the speaker.

(Otherwise the 'Why' is 'academic'.) The specimen dialogue gives us several examples of argument in a circle, and we might look for a rule which would outlaw these. The simplest possible such argument is 'Why A? *Statements A, $A \supset A$*'; and, if S and T are statements equivalent by definition, another is 'Why S? *Statement T. Why T? Statement S*'. An unnecessarily strong rule is the following:

The answer to 'Why S?', if it is not 'Statement $-S$' or 'No commitment S', must be in terms of statements that are already commitments of both speaker and hearer.

This achieves the object of outlawing circular reasoning but makes it impossible to develop an argument more than one step at a time; that is, a participant cannot make, and succeed in justifying, any statement which cannot be deduced in one step from statements his opponent has already conceded.

Let us try another tack. Even if circles appear in an argument, we might regard it as a satisfactory one provided it is capable of being restated without them. Thus we might regard an extended argument developed in answer to a string of 'Why' locutions as a satisfactory one provided the *ultimate* premisses are commitments of both parties; and, in one important particular case, if they are axioms. Now one easy way of ensuring 'goodness' of all the extended arguments in the system is to stipulate that participants always rigorously question the credentials of one another's statements with 'Why' locutions, and never concede anything except as the result of argument. There are two possible formulations which recommend themselves: the stronger is

> *If there are commitments S, T, ..., X of one participant that are not those of the other, the second will, on any occasion on which he is not under compulsion to give some other locution, give 'Why S?' or 'Why T?' or ... or 'Why X?'*

This enjoins extreme fractiousness of the participants, and can be weakened in the first clause to read:

> *If one participant has uttered 'Statement S' or 'Statements S, T' or 'Statements T, S' and remains committed to S, and the other has been and remains uncommitted to it, the second will (etc.).*

Both rules are rather drastic in effect since they imply that the process of questioning with 'Why' *cannot* stop until it reaches premisses *a priori* agreed between the two participants; and they provide no guarantee that such premisses will ever be reached. Nevertheless it will be the case that no argument which does terminate can be circular when seen as a whole, and we have succeeded in banning the Fallacy of Begging the Question.

These rules present arguments as *ex concesso*; and there are other kinds of argument. One participant may feed the other with 'Why' locutions without admitting any step of the argument himself. We have, however, allowed for this possibility in allowing forms such as 'Statement S. *No commitment S; why S?* Statement T. *No commitment T; why T?* ...', where T is equivalent by definition to S, or forms such as 'Statement S. *No commitment S; why S?* Statements T, $T \supset S$. *No commitment T, $T \supset S$; why T?* ...' and so on; namely, forms in which a 'Why' locution is accompanied by a 'No commitment' one. Black will not now be committed at any stage of the argument, yet still demands that White produce the argument. Our rules do not need to be altered to deal with this version of argument.

What other relevance does the system have to the analysis of traditional fallacies? The Fallacy of Misconception of Refutation is committed, according to modern accounts, when someone undertakes to prove one thing and proves another instead. There is no close analogy in this system of 'undertaking to prove' a thesis, but a participant who answers a 'Why S?' with a 'Statement' or 'Statements' locution (other than 'Statement $-S$') could be held to have undertaken to prove S. The form 'Statement S. *Why S*? Statements T, $T \supset U$', which is already banned by syntactical rules, might serve as a generic representation of this Fallacy, though it misses the spirit of most of the examples that are given. The Boethian version of the Fallacy does not turn at all closely on dialectical considerations.

A formal Fallacy such as (the modern version of) Consequent can, of course, be represented, as can any feature of Formal Logic. Its most explicit representation would be in the syntactically illegal 'Statement A. *Why A*? Statements B, $A \supset B$'. There is no danger here, as there is in some formulations, of confusion with argument from example.

This might be the place for a comment on the ambiguity of 'Why'. This is at least triple. In the present formulation, 'Why S?' is clearly a request for a deduction or proof of S, and a proof is a special kind of *justification* of an act of statement-making, subject to certain provisions such as that the occasion is one that is appropriate to meticulous truth-telling. Seen more generally, however, 'Why' can ask for a justification of a given statement-act independently of these assumptions: it can shade off into 'Why did you say that?', which can be answered 'In order to impress X', 'Because it seemed the right thing to say', 'For practice' or 'Because I had a fit of absent-mindedness'. These answers are so different from the previous kind that it seems appropriate to regard 'Why' as having two different possible meanings; though it is not so clear that some kinds of answer (such as inductive justifications) traditionally classified with the first are not more properly classified with the second.

A third meaning of 'Why' makes it a request for a causal or teleological explanation: 'It's cold in here. *Why?* Because the heating system is off', or 'John is in the library. *Why?* Because he has an essay to finish'. We have noticed that the Fallacy of Non-Cause as Cause originally referred to failures of deductive proof rather than failures of causal explanation, and this confusion would be more natural in a system of Dialectic that incorporated the ambiguity of 'Why' than in other kinds of analysis. Hence our system has some explanatory power that is independent of its logical merits. An analysis of Non-Cause as Cause in its original sense requires a dialectical system incorporating *reductio* reasoning, and is simpler in a system of a slightly different kind.

The concept of Burden of Proof is replaced in this system by the somewhat simpler concept of *initiative*. Generally, as in the Law of Evidence,[1] 'He who asserts must prove', in that 'Statement S. *Why S?* clearly puts the onus on the first speaker. The cruder attempts to shift the burden, such as in 'Why S? *Why —S?*', are outlawed by syntactical rules. Consider, however, the piece of dialogue 'Statement S. *Statement T*. No commitment S; why S?'. This is not at all illegal, for '*Statement T*', in failing to repudiate S, implicitly concedes it; and the first speaker, who is quite entitled to change his mind and repudiate S himself, may then question the concession. This is quite sophistical and points the need for a special rule. One approach would be to regard commitments which are the result of concession rather than of personal statement as different in character from the others, and mark the difference in the commitment-store entries. Thus let a commitment-store contain 'Sc' rather than 'S' when S is conceded rather than stated: we might add to the commitment-rules.

> *When S is written into a commitment-store it is written 'Sc' if the other participant's commitment-store already contains 'S' or 'Sc'; otherwise it is written 'S'.*

We might then interpret the rules above to stipulate that 'Why S?' cannot be addressed to a participant whose commitment-store contains merely 'Sc', not 'S'. This is an *ad hoc* solution of a

[1] Quoted by Sidgwick, *Fallacies*, p. 150, from Sir James Stephen's *Digest of the Law of Evidence*, 3rd edn., p. 100.

kind we might disapprove of on the grounds that it makes life too easy for the man who passively makes concessions. An alternative way out of 'Statement S. *Statement T*. No commitment S; why S?' would be to guarantee that when one participant gives 'No commitment S' the other always, at least, has an opportunity to do likewise before being confronted with a 'Why' locution. In fact the present system is not impossibly unfair in this respect since 'No commitment S; why S?' can always be followed by '*No commitment S*'. Our rules err, if at all, on the side of allowing potential sophists too many easy escape routes on either side of the argument.

The 'discretionary' rules we have been discussing vary in force and it is not to our purpose to make a once-for-all selection of them. Various particular systems of differing character can be obtained by making different selections. I make, moreover, no claim to completeness of this discussion: many dialectical infelicities remain as systematic possibilities even when the strongest of our rules are enforced, and it is not even certain that it is possible to legislate for good sense by purely dialectical means. The system as outlined may, however, serve as a demonstration of how much can be achieved with comparatively meagre resources.

Emphasis on words or phrases sometimes has a straightforward meaning analysable in terms of commitment. Let us suppose we have a language built on just four elementary statements 'John likes tea', 'Molly likes tea', 'John likes coffee', and 'Molly likes coffee', but with the possibility of emphasis on the nouns 'John', 'Molly', 'tea', and 'coffee'. The simplest kind of emphasis, to be called *italic* emphasis and indicated by italics, is that indicated in speech by the tone '$\sqrt{}$', as in 'John likes *tea*' when this means 'John likes tea, whether or not he likes coffee', or in '*Molly* likes coffee' when this means 'Molly likes coffee, whether or not John does'. Commitment to 'John likes *tea*' is therefore equivalent to commitment to the unemphasized 'John likes tea' together with *non-commitment* to 'John likes coffee', and similarly. It is in order to couple an italically emphasized statement with the denial of its counterpart, as in 'John likes *tea* but he doesn't like coffee', but not with the affirmation; and 'John likes *tea*, and he likes *coffee*' is a curiously inept formulation resembling a self-contradiction, which we would use in practice only if there were also some question about John's liking for, say, cocoa or orange juice. Copi's example of the change of meaning, as various words are emphasized, of 'We should not speak ill of our friends' (see p. 24 above) can be analysed in this way. But there are also other kinds of emphasis. Thus SMALL CAPITAL emphasis (tone ⌐\) indicates definiteness, as when 'John likes TEA' means 'What John likes is tea. not coffee', and carries commitment to 'John does not like coffee' as well as to 'John likes tea': similarly 'JOHN likes tea' means 'The one that likes tea is John, not Molly'. Combination of emphases gives us even more resources: '*John* likes TEA' (√_⌐\) says that John's distinct preference is tea, whatever Molly's is, and couples commitment in respect of 'John likes tea' and 'John does not like coffee' with non-commitment in respect of 'Molly likes tea'. These examples suggest that there may be other features of stress and tone with easily analysed dialectical force, though it is also possible that speakers of other languages than English may arrange matters differently. The remark should be added that these kinds of emphasis are not to be confused with the larger kinds that involve whole locutions, arguments, points or theses.

It would be instructive now to conduct some kind of formalization of the conventions of Greek public debate. For various reasons this is a project of great difficulty and I shall not attempt it in detail. Some remarks and comments are, however, in order; particularly with reference to the Fallacies that may be exemplified within it.

The 'Greek game' of Plato's earlier dialogues resembles the Obligation game (which derives from it) in having, in any given phase, two participants with specialized roles, the 'questioner' and the 'answerer'. Also as in the Obligation game, the questioner is 'Socratic' in a sense we can now make definite: his locutions carry no commitment and he has no commitment-store. When Plato has Thrasymachus complain, in the *Republic* (337)

> Socrates will do as he always does – refuse to answer himself, but take and pull to pieces the answer of someone else.

he is objecting to the practice of having one participant make no

FORMAL DIALECTIC

assertions but merely ask questions of the other; and when Socrates replies

> ... how can any one answer who knows, and says that he knows, just nothing; . . . ?

he is putting up a not-entirely-sincere justification of a procedure which Plato elsewhere holds to be appropriate to the search for philosophical truth.[1] The questioner, nominally at least, has no thesis of his own and accepts, within the limits of logic, whatever the answerer tells him.

Only if we attempt to construct a rationale for the arrangement of the *questions* might we be inclined to alter this view. In fact Socrates emerges, in the long run, as someone with strong views of his own. Biased or loaded questions are not infrequent; and something might be deduced from the way in which questions are phrased in expectation of one answer rather than another. Socrates frequently says such things as 'But our recovery of knowledge from within ourselves, is not this what we call reminiscence?' or 'Then, Thrasymachus, do you actually think that the unjust are sagacious and good?'; and he cross-questions with zeal when he disagrees, but desists when his answerer gives the answer he wants. These tactics are not necessarily disallowed by the rules of the game, but they introduce another dimension into it. In order to formulate the rules we would first have to elaborate a theory of interrogative commitments and their interaction with indicative ones, and this would take us out of our way. There are, in effect, two Greek games, of differing degrees of sophistication; and we confine ourselves here to the simpler one.

Questions must be 'definite' or 'leading' ones with small numbers of possible answers, so that our model 'Question S, T, . . ., X?' represents them reasonably well. Since the aim of the questioner is to refute the answerer he must be equipped with means of bringing about this refutation, and some such locution as 'Resolve S' must be permitted to him; but he need have no other form of locution besides these two.

The answerer's commitment-store is initially empty and his locutions are all single statements of the form 'Statement S'

[1] See Robinson, *Plato's Earlier Dialectic*, p. 80.

except that, in answer to 'Resolve S' he must say 'Withdraw S' or 'Withdraw —S', and that he may say 'Don't know S, T, ..., X'. The point of distinguishing our previous 'No commitment' locutions into these two categories will appear in a moment.

A feature of the game that it is difficult to formalize is its dependence on the views of the onlookers or 'the majority'. These people are the ultimate arbiters both concerning the admissibility of a given statement or argument and in judging the final outcome. We could, perhaps, try the following formulation: There is given, in connection with the game, a list of statements representing 'popular beliefs'. The shape of the possible plays of the game is critically dependent on this list, which is (generally understood to be) large and unsystematic. It need not be consistent. Its function is to provide prima facie answers for the answerer. Thus when the questioner asks 'Question S, T, ..., X?', and one or more of S, T, ..., X are 'popular beliefs', the answerer gives one of these as his answer; at least when this would not be inconsistent with anything in his commitment-store.

The answerer starts play with a locution 'Statement T', where T is the *thesis*, and play ends if the answerer gives 'Withdraw T'. Unless T is self-contradictory he cannot be compelled to do this – the system is semantically unforced – but he may be trapped into doing so. There must be rules, that is, requiring him to concede consequences of his various admissions. Again, this is difficult to formalize realistically. Any question, we might say, such that just one possible answer W is a consequence of the answerer's previous statements by definition or *modus ponens* or syllogism must be answered 'Statement W'; with corresponding provisions where two or more answers are so implied. But Plato also often uses argument from example or induction, and sometimes arguments from the authority of poets and others, and these are expected to dispose the answerer towards appropriate admissions though reserving to him the possibility of resistance, at least until evidence becomes overwhelming.

A special rule of the Greek game is that *every question must be followed through*; and this involves the postulation of some kind of store of 'unanswered questions', so that the game cannot properly end while this is non-empty. Hence we must distinguish a 'Don't know' answer from a 'Withdraw'. The rules for the answering of a question will be something as follows: To 'Question S, T, ..., X?', if any of 'Statement S' or ... or 'Statement X', or alternatively 'Statement —(SvTv ... vX)', is forced by an inference-rule, the answerer must give one such; otherwise if any is a popular belief he gives one such; or otherwise he may give any one; or if no one of them is in his commitment-store, he may give 'Don't know SvTv ... vX' or 'Don't know S, T, ..., X'. If the last of these is given it is *itself* placed in the commitment-store, placing the question, as it were, on the notice-paper. A 'Don't know' locution placed in the commitment-store in this way is removed only when the corresponding question is re-asked and properly answered.

It does not need saying that all the Fallacies of Aristotle's list can be realized within this framework in some shape or form; though we shall suspend consideration of Fallacies Dependent on Language until the next chapter. There is nothing much to add about some of the others, but we should notice now that since it is the questioner's task to disprove what the answerer says, there is an implicit burden of proof which makes possible the realization of Begging the Question in its full sense. The questioner commits the Fallacy of Begging the Question when he asks a question one of whose alternatives is the negation of the Thesis T, or some statement which is the last link in a deduction that would result in the disproof of T, or some relevant statement which (in the required Aristotelian sense) is less certain than the negation of T. There could be, in principle, a precise rule forbidding such questions.

The questioner may also implicitly employ *reductio ad impossibile* reasoning and even, literally, *reductio ad absurdum*, if what is 'absurd' is what is contrary to strong majority opinion. This means that the classical Fallacy of Non-Cause as Cause can be realized, and that a rule prohibiting it could be formulated.

How often may an answerer change his mind? And how long may he hold out against an inductive consequence of popular beliefs? We cannot formulate precise rules regarding these matters (and some others) unless by arbitrary stipulation. The ultimate weakness of Aristotle's own attempts to formulate precise rules is his reliance on the 'majority opinion' to enforce them, since what the majority cares to enforce is, at the very least, a contingent matter.

The systems so far considered have all turned around deductive justification of theses. It is instructive to conclude by considering one based primarily on induction from 'empirical' evidence. The game to be described is quite unambitious and designed purely to make the point of principle that an analysis of inductive procedures is possible.

The two participants are assumed to have access to a stock of empirical fact, namely, knowledge of existence of various objects characterized by sets of properties. For example, it may be known that there exists a large red chair, a red table of unknown size, a not-large not-beautiful bookcase of unspecified colour, and so on. Some of the facts may be public, some initially private to individual participants.

Each locution is of one of the following kinds:

[1] A *generalization* of the form 'All things are As' or 'All As are Bs', where A and B are affirmative or negative terms. A generalization when made is 'tabled', and must not be equivalent to any already tabled.

[2] A *denial* of an already-tabled generalization. Unless successfully challenged, a denial replaces the generalization it denies.

[3] A *challenge* to a tabled generalization or denial. A challenge is counted successful unless met as in [4] or [5].

[4] An *exemplification* of a generalization. 'All things are As' is exemplified by giving an example of a thing which is an A; and 'All As are Bs' is exemplified by giving an example of a thing which is both an A and a B or, in view of the equivalence to 'All non-Bs are non-As', a thing which is both a non-A and a non-B.[1] The example may be public or (hitherto) private.

[5] A *proof* of a generalization or denial, by deduction from tabled generalizations or denials, and/or by the giving of one or more examples as in [4].

---

[1] Cf. the 'paradox of the ravens', in an extensive literature starting with Hempel, 'Studies in the Logic of Confirmation'. My rule is not very satisfactory at this point, but no simple rule will be any better and it would take us too far off course to formulate one that is.

[6] A *concession*, or empty move. Successive empty moves, one from each participant, end the dialogue.

All examples and proofs must be logically valid: that is, any question of validity is resolved outside the dialogue proper. The same applies to objection to a generalization on the grounds of its equivalence to an already-tabled one.

A proof by combined deduction and example, as in [5], can occur in the following form. If 'All As are Bs' is already tabled, the denial of 'All Bs are Cs' may be proved – even though there is no known example of a B which is not a C – by giving an example of an A which is not a C (but of which it is not strictly known whether it is a B or not). Similarly, if 'All As are Bs' is tabled, the denial of 'All non-As are Bs' may be proved by producing a simple example of a non-B.

'Good' dialogue is that which sets up as many generalizations as possible, subject to consistency and to rigorous testing. We can imagine the process as a competitive or as a co-operative one, as we please.

If a generalization is tabled and not denied in spite of known empirical support of the possible denial, this is a case of selection or suppression of evidence, of the kind often listed as one of the abuses of Scientific Method. When a generalization is put forward of which there is no known exemplification, and it is not challenged, the result is a species of *argumentum ad ignorantiam*.

As the possible locutions have been described, a good deal depends, in a competitive dialogue in which generalizations are credited to the participant putting them forward, on initiative and tactics; and, under these circumstances, a generalization will often be rebutted by reference to a previous one which is no more firmly supported than itself. When this happens, a version of the (classical) Fallacy of Non-Cause has been committed. In certain versions of the system it would be desirable to introduce a modification to the rules to avoid this contingency.

A participant who succeeds in tabling all of the six related statements 'All As are Bs', 'All Bs are As', 'All As are Cs', 'All Cs are As', 'All Bs are Cs', and 'All Cs are Bs' before having any of them denied would seem to be proof against challenge to any of them since, whichever one is challenged, there are always two others from which it can be proved deductively; but this would be an example, of course, of Arguing in a Circle. Actually, as the rules are stated, they generally avoid this case since empirically-supported denials are given preference over deductive proofs. It could consequently occur only if there were no evidence to support any denial.

The system could form the basis of an actual game, in which the empirical evidence was represented by cards, and generalizations received scores.[1] An important feature in securing realism is sufficient complexity of the empirical evidence to discourage participants from attempting to survey it exhaustively. This is, however, easy to achieve and difficult to avoid. (Compare chess, in which the logically perfect game is beyond the capacity of the largest computer.)

The system could be elaborated in various directions to model other features of inductive reasoning and its abuses.

[1] I am indebted to V. H. Dudman for the use of some hours of his time helping me check out a trial version of such a game.

CHAPTER 9

Equivocation

Where do dialectical rules derive their authority, and who enforces them? The answer to these questions is simple, if a little disquieting in its ultimate implications. Although there are special circumstances in which there may be a Chairman, a Judge, or others whose job it is to control proceedings, in ordinary discourse there is no such person. The control of each dialogue is in the hands of the participants themselves.

There is clearly room for dispute between participants over how their dialogue should be conducted, and no dialogue will be possible at all unless there is a certain minimum of procedural agreement. We cannot legislate against all the possible abuses of dialectical procedure. and there would be little point in doing so even if we could. Provided, however, disagreement is not extreme, participants will often resolve their differences by means of one of a set of procedures that are themselves characteristic of a language or culture. There are accepted forms for the lodging of objections, their debate, and their resolution. The point of order, or procedural locution, is as much a part of ordinary language as it is of the formal rules of meetings and committees. 'What shall we discuss?', 'That doesn't follow', 'Let us leave that aside for the moment', 'Proceed!', 'I don't understand', 'That is irrelevant', 'Wait, you're going too fast for me', 'It's not for me to say' – these and similar locutions contribute not to the subject or topic of the dialogue but to its shape. We could call them 'metalinguistic' if it were not that this word is too broad and invites confusion with other dialectical phenomena such as locutions in quotation and remarks on the dialogue by onlookers.

(The concept of 'metalanguage' needs rethinking in a dialectical context.) I shall call them simply *points of order*, and contrast them with *topic points*. For various reasons we cannot always make a clear distinction between the two kinds of point, but it is useful to be able to do so on occasion.

Now it should be clear that the commonest use of logic and logical terms is as an aid to the making and debating of points of order. One's primary reason for needing to say that one statement implies, contradicts, supports, generalizes or exemplifies another is that one wishes to attack or justify certain dialectical moves. Among logical terms I include, for this purpose, the names of fallacies. Not the least of the merits of a really good classification of fallacies would be that it could be used in the formulation of appropriate points of order; and, if existing classifications are not much used for this purpose, so much the worse for them. It should be made possible in principle, as Bentham wished, that the perpetrator of fallacy be greeted with 'voices in scores crying aloud "Stale! Stale! Fallacy of Authority! Fallacy of Distrust!" and so on'.

In so far as there are accepted forms for the raising of points of order it must be possible, in principle, to reduce the forms to rule. The infinite regress latent in such a move may be avoided by various means, and does not, in any case, constitute a fundamental objection so long as we properly understand that our 'rules' are not to be conceived as rigidly enforceable in practice but rather as in the category of norms or conventions. As elsewhere in Dialectic, we have a choice where to draw the line between regarding a locution as entirely outside the permitted forms, or regarding it as legal but open to a particular reply; in this case, the raising of a point of order. From one point of view, communication may be said to have broken down between participants when one of them is regularly resorting to, say, argument in a circle; but, from another, the minimum conditions of communication are satisfied so long as properly formulated objections are listened to and answered. All that follows is that there is a certain openness about any set of rules.

These remarks prepare us for an assault on the Dialectic of Aristotle's first class of Fallacies, the Fallacies Dependent on Language; for the problem of formulating sets of rules in respect of meaning-constancy and equivocation is a difficult one, and such rules as can be offered are, as we shall see, essentially rather tentative and 'open'. But first we should reflect on why it is that the discussions of the previous chapter did not touch on these Fallacies. In describing entities denoted by letters 'S', 'T', 'A', 'B' as 'statements' we were relying on twentieth-century logical tradition to give that term meaning. The denotata of the letters are 'statements' in the sense that they are two-valued, may be combined in truth-functions, concatenated in proofs, 'asserted' as theorems, and so on. But in taking over this tradition we have also used these entities as components of 'locutions' of various kinds, and some but not all of these locutions have themselves been identified as 'statements' under a criterion involving their role in dialogue. We should not try to have things both ways. If the letters and formulae of the statement calculus really represent statements this must be because of their potential or hypothetical role as locutions in a dialogue and the part they would play if so used. This, at least, is the methodological assumption we should make as we explore dialectical systems; though some people, no doubt, would prefer the converse thesis that statements in a dialogue are such because they represent entities with the logical properties of those of statement calculus.

At all events, the smooth phrase 'Let $S, T, U, \ldots$ be statements of a logical system' begs this question and can be justified only in a preliminary exposition. If we want to lay bare the foundations of Dialectic we should give the dialectical rules themselves a chance to determine what is a statement, what a question, and so on. This general idea is familiar enough from Wittgenstein.[1] I do not think, however, that it has ever been worked out in any detail. The programme is too large a one to be undertaken here but certain features of it are of fundamental importance for us.

The thesis that I shall adopt is that all properties of linguistic entities are 'dialectical', in the sense of being determinable from the broad pattern of their use. We might call this thesis the Dialectical Theory of Logical Form or, perhaps, the Dialectical

[1] The best examples of dialectical analysis are in the 'Brown Book': Wittgenstein, *Preliminary Studies for the 'Philosophical Investigations'*.

Theory of Meaning. And I should emphasize, also, that an interest in formalized dialogues does not in itself commit us to this thesis. It is perfectly possible for someone to feel that there is much to be gained from studying the combinations of symbols in dialogue – in something the way Carnap envisaged 'pragmatics' as an extension or special application of 'syntactics' and 'semantics'[1] – while holding that meanings are states of mind, or that implications are relations between objective fact, or that logical truths are unchangeable and are perceived by a special intuition. However, I think it is true that someone who holds one of these other beliefs will be unable to take a dialogue quite seriously as a logical phenomenon and will tend to regard it as, at best, a public performance which mirrors or plays out essentially private or non-immanent processes. Such a person could countenance most of our previous chapter, but will disagree with part of what I have to say in the present one. Moreover, I shall not here rehearse any of the arguments which might convert him. If anything is here done to convert such a person, it will only be in the demonstration that the task of applying this point of view to our present problems can be plausibly carried through.

Now Equivocation, if we think of the *meanings* of sentences or terms as extralinguistic entities, becomes in essence the association of a single sentence or term with two or more such entities instead of one; but an approach such as that of the previous chapter, by locating most of the properties of the locutions in propositional letters such as 'A', 'B', 'S' and 'T', smuggles in the (what I take to be) fiction that the question of meaning can be isolated from that of dialectical properties. When the letter 'S', say, is used twice or more in a given example it is by convention the case that it has the same meaning at each occurrence; but, if meanings are to be allowed to change with context, and to be determined by the extended context, the question of whether the meaning of a given symbol changes is to be answered *a posteriori* and the question should not be begged by writing in an assumption of constancy. Where there is meaning-constancy it must be possible to deduce the fact from external considerations, and where there is equivocation the same must apply.

[1] Carnap, *Introduction to Semantics*, pp. 8 ff.

What are the external criteria of meaning-constancy?

(A) The most obvious way of finding out what a speaker means is to ask him. When he says 'By S I mean X',[1] provided we are not in doubt what he means by X, this normally settles the matter and, if the context is an argument in which S might have been equivocal, we can turn our attention to the actual truth or falsity of the statements in it, with S interpreted as X. For example, a military commander can 'locate' one of his units either in the sense of finding out where it is, or in the sense of deciding where to put it; but, if a staff-officer ambiguously reports 'Battalion B has been located at grid-reference G', he can always be asked which of the possible senses he intends. We can imagine that there might be a slippery communication-situation in which someone, or some chain of people, effectively reasons:

	(1) Battalion B has been told to go to G;
that is	(2) Battalion B has been located at G;
therefore	(3) Battalion B must be at G.

Battalion B may not be at G, the enemy may strike and win the battle; but the bearer of (2), if he has survived, can be asked 'Did you mean "located" in sense such-and-such or sense so-and-so?', and all may then be made plain. Possibly, of course, no one will ever ask him, and in this case we shall never know for certain what he meant; but this is a practical point that need not concern us so long as we are sure that a suitable answer *would have been* forthcoming if the question had been put.

Let us formulate this first attempt at a dialectical meaning-criterion, then, as 'What a person means by his utterances is what he tells or would tell you he means if asked'. This is a better and more powerful criterion than it at first appears, for most of the objections that are most naturally lodged against it are quite easily answered. Consider, first, the objection that 'By S I mean X' merely defines one linguistic entity in terms of another, so that we remain imprisoned in language. This complaint arises from a too narrow conception of what it is to explain the

[1] The letters 'S' and 'X' may represent terms, whole statements or other linguistic entities; the distinction is not important to the present discussion and I shall make varying assumptions in different examples.

meaning of a word: there are well-understood means such as ostensive definition which do get outside language for us, and it is not at all unusual for the meaning of a family of terms to be made clear simply on the grounds of their mutual relations. Otherwise, no one could ever learn a language by the 'direct' method.

Secondly, it could be objected that locutions such as 'By S I mean X' occur only at a very sophisticated level of language, and that we could never come to learn to use them until we had independently learnt the meanings of large numbers of more elementary terms and idioms. The answer to this objection, however, must be straight denial; for although the meaning of the word 'mean' may perhaps be a sophisticated matter, the processes of explaining a meaning, giving a synonym, and so on, are among the most primitive processes of language. It is hardly conceivable that there should be a practically usable language that did not contain, or have associated with it, idioms permitting the explanation of meanings.

Thirdly, it might be objected that our formulation does not apply to the meanings of the linguistic entities themselves but only to 'what the speaker means'; and that language is not tied to any given one of its users or occasion of its use. There are two answers to this. The first is that the thesis as it stands is actually a little narrower than it need be, but that it can easily be broadened. The meaning of a term is, it is true, not just a matter of what a speaker or writer intends it to mean but also a matter of what a hearer or reader understands by it, what an average speaker or an average hearer would mean by it in normal circumstances, and so on; but we can determine what a hearer takes a word to mean by asking him, and we can determine what an average speaker, hearer, reader, or writer means by conducting a poll. We can, perhaps, determine what a fictional speaker or hearer means by devising an imaginary explanation-request and answer in accordance with the canons of fiction. Thus 'hearer's meaning', and the others, can be defined analogously to 'speaker's meaning'. But a second answer is appropriate to those who remain unsatisfied with this extension of the criterion and insist on asking for a definition of '*the* meaning'. It is to ask: What other kind of meaning can there be? Besides the various possible explanations of meaning that could be elicited by asking users, it is difficult to see that any other source of explanations is possible. This pure Platonism must be rejected.

Thus the criterion stands up reasonably well to some of the possible objections. Unfortunately, there are difficulties of another order. It should be clear, for one thing, that there are often cases in which we feel that what a man means is *not* what he says he means; and this implies some different test of meaning from the one we have given. More disturbingly, it sometimes happens that when a speaker is asked to clarify an ambiguous statement he is unable to do so because the confusion is more than verbal. Let us return to the example of the military commander. Staff officers and their aides are apt to conceive units as chessmen that can be moved at will, and someone whose reasoning is directed by this model will be apt, to this extent, to recognize only one sense of the word 'located'. Asked to distinguish two senses he will at first be at a loss. If so, it seems to follow that a person may sometimes not know what he means or, at least, that he may be unable to give the kind of detailed account of his meaning that is necessary to resolve questions of meaning-constancy or equivocation. In short, we shall have to abandon this criterion and look for some other one which includes an analysis, among other things, of what it is for someone's thinking to be model-dominated.

(B) If, then, we cannot rely on what a speaker *says* he means, to what extent can we determine meaning from use in (what must be called) 'zero order' contexts? If a word W is under study and we disregard its occurrence in such contexts as 'By W I mean so-and-so' and turn instead to its primary application in such contexts as (if it is a common noun) 'All (W)s are (Y)s' and 'So-and-so is a (W)', is it here that questions of meaning are resolved? (In these formulations the parentheses represent syntactical 'anti-quotes', such that when the expression within them refers to a word the expression including parentheses refers to the word's reference.)

There are two well-known answers to this question in recent philosophical literature. Both involve the recognition that some statements, more than others, tend to reflect the meanings of the words in them in the sense that their truth or falsity is relatively little dependent on empirical or extralinguistic fact. Quine ("Two

Dogmas of Empiricism'), though he denies that there is a coherent distinction to be drawn between 'analytic' and 'synthetic' statements, erects, in its place a distinction between a higher or lower degree of 'corrigibility' or 'immunity to revision'.[1] Given a body of beliefs, and an experience in some way at variance with them, we always have a choice of different ways of adjusting our beliefs to accommodate the experience; and in making this choice we give favouritism to certain beliefs and prefer, if possible, to adjust certain others. The favoured, or relatively 'incorrigible', beliefs include the higher-order best-entrenched theoretical ones, particularly those of Mathematics and Logic. The less-favoured or 'corrigible' ones are the ones whose reversal creates no great strain in the system of beliefs as a whole, such as those of ill-confirmed immediate experience.

Grice and Strawson ('In Defense of a Dogma'), though apparently defending the distinction between analyticity and syntheticity against Quine's attack, given an account of it which also makes it a relative matter. This time, however, it is not the entrenchment or corrigibility of statements that is emphasized but rather the comprehensibility of their negations. The man who says 'My neighbour's three-year-old son does not yet understand Russell's theory of types' says something synthetic since, however surprised we would be to find it false, we can still give some meaning to the hypothesis that it should be so; but 'My neighbour's three-year-old child is not yet an adult' is such that, if someone were to deny it, we would find his statement incomprehensible. 'You mean that he is unusually well-developed?' we would say, or 'You mean that he has stopped growing?'; but if glosses of this kind were all rejected we would have no option but to regard the statement as meaningless altogether. Grice and Strawson do not intend their account as a complete one but, for what it is worth, it reduces analyticity to a behaviour-pattern of the hearer in reaction to an actual or hypothetical denial, and is generically similar to Quine's. Both accounts are 'dialectical', in that they refer their respective explications of analyticity or incorrigibility to patterns of verbal behaviour. Quine, it is true, thinks in terms of an average or corporate behaviour of modern scientists, and Grice and Strawson think rather of individual idiosyncracies; but in both cases it is clear that questions of meaning are to be resolved into questions of analyticity or incorrigibility of verbal formulations, and these, in turn, into behaviour patterns.

Expanded to meet our present preoccupations, the account must go something like this: Meanings of words are, of course, always relative to a language-user or a group G of language-users. The meanings any group G attaches to words are determinate in terms of the (zero-order) statements that are relatively incorrigible or analytic to G. Two words W and X have the same meaning for G if members of G, in the face of any experience, consistently and stubbornly allocate the same truth-value to any zero-order statement containing X as they do to the corresponding statement containing W and, in particular, if they consistently and stubbornly maintain the truth of 'All (W)s are (X)s and all (X)s are (W)s'; and, perhaps (though this is independent), express puzzlement when faced with statements such as 'Not all (W)s are (X)s' or 'This A is an (X) but not a (W)'. The words W and X have different meanings for G if there is a zero-order statement containing X to which they are quite prepared to allocate a truth-value different from the corresponding one containing W (etc.).

There is a reverse side to this doctrine, which needs to go as follows: Since the language-behaviour of some person or group may be unsystematic or incoherent, it is not necessarily the case that questions of meaning are resoluble. If some members of a group assent to, and others dissent from, certain statements it may not be possible to say, for that group, either that W and X are synonymous or that they are not. Furthermore, the statements even of an individual may be mutually 'inconsistent', in a primitive sense of that word, namely, that they do not fit together into a pattern. It is only in so far as a regular pattern of use can be determined that it is possible to make suitable judgements about meaning.

Now let us turn to equivocation. What, in the zero-order uses of a word W by a group G, could lead us to judge that W is equivocal? The question is not easily answered, and we must subdivide it a little. Let us start by asking how we could determine that a word W has two or more *different* meanings. Some obvious tests present themselves: chairs and questions can both be 'hard', but it is easy enough for us to sort out two meanings on the grounds of categorial difference using, if necessary, the refined methods of Sommers ('The Ordinary Language Tree' and 'Types and Ontology'). Even when no relevant categorial difference presents itself as, perhaps, in the case of 'bank' in 'I'm going down to the bank', it may be the case that the pattern of use splits sharply into two sub-patterns: the occasions on which people talk about financial institutions and the things they say about them seldom overlap with the occasions of their references to, or their statements about, river-verges. In such cases the distinguishing of two meanings is easy but seldom necessary and hence in a certain sense gratuitous.

Equivocation differs from double-meaning because, first, we must assume the existence of an invalid argument based on meaning-shift and, secondly, because we must assume that the perpetrators of the argument either deceive themselves, or set out to deceive other people, into thinking the argument valid. But equivocation, as we remarked in chapter 1, may be of two kinds, gross and subtle. It is, presumably, a sufficient account of the gross kind that the double-meaning involved should be as just described. Since it is only *per stupiditatem* that anyone is ever deceived by them, equivocations of this sort are of little interest to the logician.

The subtle variety is a different matter. Let us revive the example we used in chapter 1. Someone says 'Ignorance of the law is no excuse', and says it disapprovingly, in such a way as to leave no doubt that he regards ignorance as morally blameworthy. We may suppose, in fact, that he regularly and systematically fails or refuses to make a distinction between legal and moral obligations, and accepts the consequences. His reasoning contains arguments that others consider equivocal; for example

All acts prescribed by law are obligatory.
Non-performance of an obligatory act is to be disapproved.
Therefore, non-performance of an act prescribed by law is to be disapproved.

It is not clear, however, that there is or need be any feature of *his own* zero-order utterances that betrays or indicates this equivocation.

The point is a quite general one. When someone reasons, syllogistically, say, and we wish to condemn his argument as an equivocation on the middle term, our grounds for doing so will be that we consider the premisses actually or possibly true and the conclusion actually or possibly false; but these cannot be grounds which will appeal to the person who puts the argument forward, since he must be supposed to be deceived by, or out to deceive others with, the argument. If, on this theory, we suppose an argument to be capable of *deeply* deceiving someone, we thereby suppose it to be capable of creating for him a whole pattern of use of the words involved in it; and thereby destroy the supposition that the words are, for him, equivocal.

Our conclusion could be that the theory that a man's zero-order use of language determines what he means by his words is untenable. But it could also be that there is no such thing as a deep or subtle equivocation. Before exploring further, let us see what other dialectical tests of meaning-constancy are possible.

(C) Sometimes cases arise in which we want to say that someone is deceived by an argument 'temporarily', or 'against his better judgement', and that later or quieter reflection leads him to a reappraisal in which he sees the fault; perhaps, an equivocation of the middle term.

This kind of case needs to be mentioned but, as a test for equivocation, it is really outside our terms of reference. The temporary nature of his assent to the conclusion is not in itself an argument for its invalidity; the fact that his repudiation of it was later, or quieter, is not an argument for its correctness; and we only beg the question if we refer to the argument as a 'deception' or say that he later 'sees the fault'. We have not yet shown that these terms have a dialectical analysis.

In fact, the whole question of a dialectical theory of *truth* and *falsity* might be said to be open. The simplest of all demonstrations that an argument is invalid – and hence, perhaps, equivocal – is the demonstration that its premisses are true and its conclusion false. Dialectical considerations, however, do not provide such tests since they do not provide criteria of truth or falsity for more than a very restricted class of the statements we make.

If there is a dialectical theory of truth it must run something as follows: 'It is true that S' means (very nearly) the same as 'S', and 'That is true', 'That is false' are phrases used in dialogues to indicate agreement and disagreement. 'Is it true that S?' is (very closely) the same question as 'S'?, and so on. When the abstract nouns 'truth' and 'falsity' are used they are translatable, not always very directly, into more elementary terms, and hence dissoluble into locutions relevant to actual or projected cases of agreement or disagreement. 'S and T have the same truth-value' means (approximately) '(I agree to) T if and only if (I agree to) S'.

Now the assumption that the premisses of some projected argument are true and the conclusion false is an assumption *we* non-participants make and which cannot commit the participants. If, on the other hand, our hypothetical arguer who trusts his better judgement comes to this conclusion about truth-values, this says nothing except that his use of words, in so far as it is a quiet and reflective one, is not such as to lead him to regard the argument as valid, even though his use in the heat of the moment may have been different.

(D) Whatever theory is adopted, it must explain a certain asymmetry between 'Yes' and 'No' answers to questions of meaning-constancy. Although there is no contradiction in supposing of quite ordinary words in quite ordinary contexts that they are equivocal and misleading, we almost never suppose any word to be equivocal until we get into trouble with it. When two people disagree over some kind of inference, charges of equivocation may begin to flow; but, when we are in agreement and think that there is nothing to disagree about, to envisage simultaneously that there is or might be a hidden equivocation is inappropriate for us *because it is unnecessary*. The onus of proof is on the side of the person who makes such an assumption, and the assumption is equivalent to an assumption of the *possibility* of disagreement.

Consequently, a theory of meaning-constancy that can be applied independently of whether the constancy actually matters or makes a difference must be fundamentally on the wrong tack. There is, as we might put it, a *presumption* of meaning-constancy in the absence of evidence to the contrary. The presumption is a methodological one of the same character as the legal presumption that an accused man is innocent in the absence of proof of guilt, or that a witness is telling the truth: it is not, of course, itself in the category of a reason or argument supporting the thesis of meaning-constancy, and least of all is it an argument for the *impossibility* of equivocation. Dialectic, however, has many presumptions of this kind, whose existence is related to the necessary conditions of meaningful or useful discourse. It is a presumption of any dialogue that its participants are sober, conscious, speak deliberately, know the language, mean what they say and tell the truth, that when they ask questions they want answers, and so on.[1]

Now the presumption of meaning-constancy – and we might say that this presumption applies with especial force in the case of two adjacent uses of a word in the statement of an argument – is not inconsistent with the doctrine that the meaning of a word is the pattern of its use, or even with the doctrine that a man's meaning is what he *says* it is; but it gives any such doctrine a new character and a special twist. We may have to say, for example, that in so far as there is a presumption that W is constant in meaning there is a presumption that any given use of W is part of a pattern, or that the user's explanations of his meaning are mutually coherent. And if a given use of W by person P seems not to be part of a pattern of use by P, it is always possible that there is a *developing* pattern and that a retrospective assessment may alter the verdict. In conjunction with these other doctrines the presumption of meaning-constancy implies that an apparent equivocation always carries with it a presumption that there is an actual or developing meaning of the locutions in question, of such a character as to render the suspect locution non-equivocal.

This makes the formal logician's assumption of the meaning-constancy of his symbols at once both more and less reasonable; more, because his assumption is our presumption; less, because it assumes that meaning-constancy is axiomatically and absolutely achievable.

When we reflect that charges of equivocation are themselves dialectical locutions, and that they are procedural in nature – that is, points of order – we may also be led to wonder whether a

[1] Jurisprudence recognizes three kinds of presumptions, of which these resemble the *juris sed non de jure*, presumptions 'at law but not with legal force'. See *Kenny's Outlines of Criminal Law*, § 489.

better analysis of them might not be given in terms of their procedural role, resting on features not realizable at the topical level. Points of order are our means of pressing an opponent to adopt this or that procedure or to give the dialogue this or that shape, and they may also be our means of pressing others to adopt certain uses of words rather than others. We perhaps assume too readily that a point of order, when it raises an objection to some locution, stigmatizes a prior fault. Though it may be appropriate to say so of an objection to a gross equivocation, the subtler variety should sometimes be regarded less as a fallacy than as an idiom in disuse.

There are two ways in which 'points of order' might be incorporated, at an elementary level, into a formal, dialectical language. The first would be to treat each *kind* of point of order on its merits, as a special locution with prescribed functions; as, in committee procedure, there is one, stylized form of closure, 'I move that the motion be now put'. The other would go to the other extreme and provide a procedural metalanguage with all or most of the facilities of the object language but including also means of referring to locutions of the object language and dissecting their relevant properties. Neither is quite realistic. Procedural locutions do not, as in the first model, lack a grammar; but neither are they, as in the second, to be regarded as debatable in quite the same sense as topical ones. The consideration of points of order is essentially perfunctory. There is something wrong with a dialogue in which points of order are endlessly debated, or in which they are subject without restriction to yet further points of order.

We should now remind ourselves that much of what we have been saying is implied in Sextus's criticism of the concept of fallacy. For Sextus, even proper names do not have meanings independent of the context of their use, as he makes clear with the example of the two servants called Manes. The order 'Fetch Manes' is unambiguous if it is given when only one of the two servants is on duty, but otherwise the person to whom the order is given will have to ask 'Which one?'. In the same way, the order 'Bring me some wine' can be carried out without question if there is only one kind of wine available, but needs to be re-referred back when uttered by a man rich enough to have two.

The name 'Manes' and the phrase 'some wine' are, in themselves, neither ambiguous nor unambiguous and neither, we must assume, is any other word or phrase. What matters is that locutions involving them should play their appropriate, or demanded role; and, provided they all do this, it makes no sense to explore their meaning any more finely.

So much, at least, Sextus says about the question of what a word means on a particular occasion of its use. But it is a reasonable guess that this is also what he would say about the question of *the* meaning of a word: words have no meanings apart from the meanings they bear on particular occasions and though, if a pattern of use develops, we may describe the pattern and discover an average or norm this norm is no more than the expression of the demands for satisfactory communication on the individual occasions from which it is derived.

Sextus, however, goes much further than we have so far done, in that he makes, or seems to make, a criticism not merely of the doctrine that terms can be clearly equivocal or unequivocal but also of the whole concept of a fallacy. To be specific, he makes criticisms which, if they are to be sustained, can be interpreted as affirming the dialectical character not merely of the meanings of terms, but of logical form itself.

Let us consider further his example, quoted in chapter 3, to prove the uselessness of a logical doctrine of fallacies:

> In diseases, at the stages of abatement, a varied diet and wine are to be approved.
> But in every type of disease an abatement inevitably occurs before the first third day.
> It is necessary, therefore, to take for the most part a varied diet and wine before the first third day.

The argument, he tells us, is valid so far as the logician is able to tell; for only the physician with his special knowledge will be able to see that the word 'abatement' is equivocal and refers, in one case, to the general abatement of the disease and, in the other, to the periodic troughs in the fever-cycle. The physician sees this *because he knows the conclusion to be false*. It is not totally anachronistic to conceive the 'special knowledge' of the physician as a knowledge of empirical fact, unavailable to the logician since he deals only with the *a priori*. In modern terms, then, what Sextus is saying is that the inferences licensed by the logician carry no authority except what they derive from their conformity to our independently-derived preconceptions of the truth and falsity of the statements occurring in them.

> Thus, suppose there were a road leading up to a chasm, we do not push ourselves into the chasm just because there is a road leading to it but we avoid the road because of the chasm.

Properly understood, this says something about the whole status of the logic of inference. We can underline Sextus's thesis by directly raising the question of the application of a logical inference-schema. Let me suppose that I accept a certain premiss or premisses P and a certain inference-process from the application of which I should be led to deduce conclusion Q, but that I have no independent belief either way regarding the truth of Q. Should I accept Q? The question seems to be without sense: assent to P and assent to the inference-process *implies* assent to Q. But does it? It depends on what is meant, precisely, by 'assent to the inference-process'. There is a possible sense of this phrase such that it is not possible for someone to assent to premisses and inference-process without assenting to the conclusion, or such that a man who accepts premisses but not the conclusion cannot properly be said to have accepted the process. However, acceptance of an inference-process may also be 'formal' only, and it is in this sense that we customarily accept the *schemata* put before us in Logic-books. Having put up their *schemata*, the writers of Logic-books are sometimes only too conscious that they cannot, as it were, hand out an unconditional guarantee with them: there must be a saving-clause against improper use. A great deal hangs, then, on the question of what use is 'proper' and what not.

That Formal Logic cannot formalize its own application needs no argument: it takes an enterprise of a different order to do that. But the point that Sextus the sceptic feels bound to make for us is that this new enterprise cannot hand out guarantees either. At least sometimes – and, Sextus perhaps thinks, always – the discovery that a given inference is invalid is made *a posteriori*, from independent knowledge of the falsity of the conclusion.

The problem is particularly pressing in the case of 'fallacious' arguments since, by definition, they are arguments that *seem valid*. In the grosser kind of fallacy the deception may be a gross one, which is easily dissipated and seen for what it is; but in subtler fallacies it may not be possible to be sure of the source. In fact, if the test of validity is in any way independent of the actual truth and falsity of premisses and conclusion, the subtle fallacy shades into the valid inference, with no possibility of drawing a dividing line.

Let us sharpen up our example and suppose that, from premisses P which I accept, by a process I accept, I am led to a conclusion Q that I *reject:* What am I to say? When I am asked whether I think Q to be true, the only answer I can conscientiously give is 'No', and to say anything else would be dishonest. Similarly, if I am asked whether I think P to be true, I must say 'Yes', for to give any other answer would be dishonest too. What, now, if I am asked 'But doesn't P imply Q?'? I shall be in trouble if I merely answer 'Yes' to this, but it would be possible for me to say something like: 'It seems to me that the inference from P to Q is valid. Assuming that I am right in thinking P true and Q false there must be something wrong with this inference, but I am unable to say what it is.' What might it be? If our interests are logical we shall, no doubt, examine the details of the inference closely: check the formal inference in accordance with recognized logical systems, double-check our formalization, run through lists of fallacies to see if they give us any clue to the failure. But what if all reasonable efforts fail? One reaction might be to label the inference a 'paradox' and regard it as a 'difficulty' for the logical system within which it is most conveniently analysed, but this would concede the *a posteriori* nature of logical investigation. It is difficult to conceive of any other reasonable reaction apart from simple suspense of judgement.

What needs to be emphasized is that logic is at least partly a normative enterprise and seeks to impose a certain order on our possibly recalcitrant personal predilections. It is, moreover, possible to conceive even rules such as *modus ponens* primarily as rules of dialectical procedure. There is, we might say, no contradiction in a man's *thinking* P to be true, Q to be false and P to imply Q, and this is our occasional fortunate or unfortunate condition; but he must not *say* that this is the case. He may say it, at least, only with the prefix 'I think'.

These points apply generally in the discussion of any kind of fallacy, but it should be clear that they arise with special force when we consider the Fallacies Dependent on Language, of which we may continue to take Equivocation as typical. It is by now more than ever clear that there are many cases in which the decision to regard a word or phrase as equivocal is come to as one among several possible escape-routes from a threatening contradiction. That this may sometimes be the *only* reason for the decision is a sceptical suspicion which makes us look, now, for firm ground for our feet. There is, after all, no doubting the procedural importance of *charges* of equivocation in forcing the clarification, and perhaps creation, of shades of meaning of the words accused in them.

For how do words change their meaning? What dialectical phenomena accompany such change? There is no contradiction in supposing a new use of a word to become current overnight, by magical universal agreement; but we may be sure that this is not the usual case. Again, the history of language reveals a slow drift of meanings imperceptible in the short term; but this kind of change can be of interest only to someone who wishes, as it were, to converse across the centuries. Of far more significance is the *inventive* use of words: a model catches the attention, its properties are verbalized and, without apology, we have a new use thrust on us. The model may be useful or misleading, permanent or fleeting. If it is a bad model, we shall be in our rights in regarding arguments relying on it as equivocal; but if it is not, to do so would be at best academic and at worst obstructionist.

A charge of equivocation may be thrown back – 'You are using W in one premiss in sense J and in the other in sense K'. 'No, I am using it in both cases in sense L'. The extent to which one may *demand* that a word be used in a given sense is variable: we do not, dialectically speaking, have complete 'freedom of stipulation' in the sense in which this sometimes asserted but the justification is, again, pragmatic and no one can make decisions that *must* be binding on others. 'You are misusing the word W' is a possible form of point of order.

Our thesis might be summed up by saying that equivocation

is a procedural, non-topical concept. The question of whether a given term is or is not (subtly) equivocal in a given context or family of contexts is not one that may be decided by looking at the term itself, but rather a question of how one may best order the discourses within which the word appears. We can understand the word 'equivocal' only by seeing that it is not a descriptive term, but rather one that is used in the making of certain kinds of procedural points. 'That is equivocal' does not describe a locution, but objects to it.

A further argument for the non-topicality of equivocation rests on the existence of dialectical paradoxes resembling, though not precisely parallel with, the paradox of the Liar. The simplest of these, in fact, is a dialectical version of the Liar-paradox itself, and provides an argument for the non-topicality of the concept of truth. The man who says of *any* of his own, current, utterances that it is false presents us with a curious problem; for a necessary condition of a lie is that it should be a prima facie deception but, once a statement is *stated* to be a deception, it is so no longer. What is openly stated to be false is not, in the large view, a falsehood. 'S – no, that's false' may be comprehensible as a change of mind, equivalent to 'S – no, not-S', but 'S and that's false' is a species of nonsense.

A dialectical paradox is generated by a kind of self-reference; not, however, a reference of a given statement to itself, but a reference by a speaker to one of his own current utterances. (A 'current' utterance is an utterance to which he remains committed.) We could characterize the enterprise of the dialectical fallacy-monger by borrowing some terminology from the early Wittgenstein, and saying that he tries to *say* what cannot be said but can only be *shown*.[1] Wittgenstein, after all, indulged in this kind of paradox himself when he wrote at the end of his book (6.54) that the other statements in it were all nonsense. It is possible for me to say S and to know, and to show by some such means as uncertainty of manner, that it is false; but it is not possible to *commit* oneself both to S and to S's being false.

This being so, something equivalent will apply to 'S and that's true'. The phrase 'and that's true' cannot say more than S says, except in the sense that it may add emphasis, or increase

[1] Wittgenstein, *Tractatus Logico-Philosophicus*, 4.121 ff.

conviction, or otherwise achieve something that could have been *shown* in other ways, such as by saying '*S*' very loudly. Moreover, the second-person locutions 'What you say is false', 'What you say is true' take over some part of the paradoxicality from their first-person equivalents since, considered as topical locutions, they would have to be capable of commanding agreement and disagreement, and this would involve the hearer in committing himself to paradox in the first person. Reference by any participant to another participant's current locutions may also engender paradox. 'What you say is false', and 'What you say is true', may represent elliptical disagreement or agreement with the hearer's locutions, or they may be, or be part of, point-of-order locutions; but they cannot make topical points that essentially involve the concepts 'true' or 'false'.

That 'valid' and 'invalid' are ultimately of the same character could be argued at length, but it would be necessary first to strip from them the meanings they tend to assume within formal systems, in which they indicate topically-discussable conformity with this or that formal norm. I shall confine myself to arguing the case of 'equivocal', which has no clear formal meaning provided we exclude its grosser manifestations. What could 'I am equivocating' mean? It is of the essence of an argument, as we were at pains to insist earlier, that it be put forward in support of its conclusion: otherwise it is merely 'hypothetical'. But someone who argues, say, from premisses P to conclusion Q, and then adds that his argument is equivocal, has implicitly negated the seriousness of his purpose in supporting Q, and no longer really argues. Hence he does not even really equivocate. 'P, therefore Q, and that's equivocal' is a piece of nonsense of the same order as 'S and that's false'. The locution 'That was an equivocation' may, of course, be used to unsay an argument but it cannot be used by the arguer to say something additional. He can *show* that he is equivocating, as those who equivocate commonly do, by his hesitancy or, more probably, his dogmatism; but he cannot say it. In the same way, 'I am not equivocating' is an empty assertion except for possible subsidiary functions such as adding emphasis or rejecting an objection, and 'You are equivocating', though very much to the point as a point of order, cannot be topical either.

The distinction between topical and procedural locutions can be seen to fulfil the same role for us, in the solution of paradoxes, as theories of levels of language have done for others in connection with paradoxes of a more traditional nature. A dialectical system cannot unrestrictedly admit the predicates 'true' and 'equivocal', together with means of reference to a speaker's or hearer's current locutions, into its topical object-language.

One answer to all of this might be that the logician is never a participant in the dialogues whose locutions he studies, but always an onlooker; and that he may, consequently, make whatever comments he wishes concerning the arguments that participants put forward, without becoming involved in paradox. He can say 'X *was* equivocating when *he* said so-and-so'. He can do this impartially just as, when we stand aside from an argument between X and Y, we can say 'X was really right: what Y said was false'. Do logicians, and scientists, speak only in the third person? This is an intellectual fancy that is both comforting and delusive in its implications. It is capable of turning into the claim to be above criticism, like the armchair strategist whose failure to win battles is due to the lack of co-operation of the enemy.

Large-scale and would-be-monolithic enterprises like Logic and the other branches of learning do not superannuate Dialectic or escape its processes, for the onlooker is in an arena of his own. That the interchange may take place in book-length or article-length locutions, with a large and ill-knit group of participants, makes a great deal of difference, no doubt, to the rules; but the rules must have, in common with those that apply to dialogues more properly so-called, the feature that they apply within a variable language. No one will understand equivocation who thinks that words permanently retain their meanings; or, particularly, who thinks that it is his task to make them do so.

The road to an understanding of equivocation, then, is the understanding of *charges* of equivocation. For this, the development of a theory of charges, objections or points of order is a first essential.

BIBLIOGRAPHY

ABELARD, PETER (1079–1142). *Dialectica*, edited by L. M. de Rijk. Assen, Van Gorcum, 1956. 2 vols.
Logica Ingredientibus in *Peter Abaelards Philosophische Schriften*, edited by B. Geyer. Münster i. Wien, 1919–27. *Beiträge zur Geschichte der Philosophie und Theologie des Mittelalters*, XXI, 1–3.

ADAM OF BALSHAM (ADAM BALSAMIENSIS PARVIPONTANUS, d. 1181). *Ars Disserendi* or *Dialectica Alexandri* (1132), edited by L. Minio-Paluello. Rome, Storia e Letteratura, 1956. *Twelfth Century Logic. Texts and Studies*, I.

AGRICOLA, RODOLPHUS (ROELOF HUUSMAN, 1444–85). *De Dialectica Inventione* (1497). Edition entitled *De Inventione Dialectica Lucubrationes*, Cologne, 1539. Facsimile of this, Nieuwkoop, Graaf, 1967.

ALBERT THE GREAT, *saint* (ALBERTUS MAGNUS, 1193?–1280). *Opera quae hactenus haberi potuerunt* . . . edited by Petrus Iammius. Lyons, Prost, 1651. 21 vols. The commentary on the *Sophistical Refutations* is in volume 1.

ALDRICH, HENRY (1647–1710). *Artis Logicae Compendium* (1691), edited by H. L. Mansel as *Artis Logicae Rudimenta*, 1849. 4th ed., Oxford, Hammans, 1862.

ALEXANDER OF APHRODISIAS (2nd–3rd century). *In Aristotelis analyticorum priorum librum 1 commentarium*, edited by M. Wallies. Berlin, Reimer, 1883. *Commentaria in Aristotelem Graeca*, vol. 2, part 1.
In Aristotelis topicorum libros octo commentaria, edited by M. Wallies. Berlin, Reimer, 1881. *Commentaria in Aristotelem Graeca*, vol. 2, part 2.

(PSEUDO-)ALEXANDER (=MICHAEL EPHESIUS? 11th century?). *Alexandri quod fertur in Aristotelis Sophisticos Elenchos commentarium*, edited by M. Wallies. Berlin, Reimer, 1898. *Commentaria in Aristotelem Graeca*, vol. 2, part 3.

ÅQVIST, LENNART ERNST G:SON (1932–). *A New Approach to the Logical Theory of Interrogatives. Part 1, Analysis.* Uppsala, Filosofiska Föreningen, 1965.

ARISTOTLE (384–21/2 B.C.). *The Works of Aristotle*, Translated into English under the editorship of W. D. Ross. Oxford, Clarendon, 1908–52. 12 vols. Marginal page and line numbers correspond with the Oxford edition of Bekker's Greek text *Aristotelis Opera ex recensione Immanuelis Bekkeri*. Oxford, University Press, 1837.
  Aristotle on Fallacies or *The Sophistici Elenchi*. With a translation and commentary by Edward Poste. London, Macmillan, 1866.
  Elenchorum sophisticorum Aristotelis libri duo, An. Manl. Sev. Boetio Interprete. Paris, 1860. *Patrologiae Cursus Completus Series Latina*, edited by J.-P. Migne. vol. 64, cols. 1007–40.
  Aristotelis opera cum Averrois commentariis. Venice, 1562–74. Photoreprint of this, Frankfurt/Main, Minerva, 1962. 12 vols. in 14. Vol. 1 part 3 contains *Sophistical Refutations* and commentary.
ARNAULD, ANTOINE (1611–94). *La Logique, ou L'Art de Penser* (1662). Edition entitled *The Art of Thinking* (*Port Royal Logic*), translated and with an introduction by James Dickoff and Patricia James. New York, Bobbs–Merrill, 1964.
ARNIM, HANS FRIEDRICH AUGUST VON (IOANNES AB ARNIM, 1859–1931). *Stoicorum veterum fragmenta*. Stuttgart, Teubner, 1903–24. 4 vols.
AVERROES (IBN RUSHD, MUHAMMAD IBN AHMAD, 1126–98). *Tahafut al-tahafut* (*The incoherence of the incoherence*). Translated from the Arabic by S. van den Bergh. London, Luzac, 1954. 2 vols.
  See also ARISTOTLE.
BACON, FRANCIS, *Viscount St. Albans* (1561–1626). *Works*, collected and edited by Spedding, Ellis and Heath, London. 1858–74. Facsimile edition, Stuttgart–Bad Cannstatt, Frommann, 1961–3. 14 vols. Vol. 3 contains *The Advancement of Learning* (1605), *Temporis Partis Masculus* (Latin, c. 1603?), *Valerius Terminus* (Latin, 1603 and later corrections), *Redargutio Philosophiarium* (1608–9). Vol. 4 contains *The New Organon* (translation of *Novum Organum*, 1620). Vol. 6 contains 'Of Cunning' in *Essays*, 1612.
BACON, ROGER (1210/15–94). *Sumule dialectices* (c. 1245?) in *Opera hactenus inedita Rogeri Baconi* . . . edited by Robert Steele. Oxford, Clarendon, 1909. Fasciscule 15.
BAR-HILLEL, YEHOSHUA (1915–   ). 'More on the Fallacy of Composition.' *Mind*, 73 (1964), pp. 125–6.
BECHMANN, FRIEDEMANN (FRIDEMANNUS BECHMANNUS, 1628–1703). *De modo solvendi sophismata*. Leipzig, 1648.
BELNAP, NUEL DINSMORE, Jr. (1930–   ). *An Analysis of Questions: Preliminary Report*. Santa Monica, California, Systems Development Corporation, 1963.

BENTHAM, JEREMY (1748–1832). *The Book of Fallacies: From Unfinished Papers of Jeremy Bentham*. By A Friend. London, Hunt, 1824. Republished under the title: *The Handbook of Political Fallacies*, revised and edited by Harold A. Larrabee. Baltimore, Johns Hopkins, 1952.

Tactique des Assemblées Législatives, suivie d'un Traité des Sophismes Politiques: ouvrage extrait des manuscrits de M. Jérémie Bentham, . . . by E. Dumont. Genève, Paschoud, 1816.

BLACK, MAX (1909–). *Critical Thinking: An Introduction to Logic and Scientific Method* (1946). 2nd ed., Englewood Cliffs, N. J., Prentice-Hall, 1952.

BOCHENSKI, INNOCENTIUS M. (1902–). *A History of Formal Logic*, translated and edited by Ivo Thomas. Notre Dame, Indiana, University of Notre Dame Press, 1961.

BOETHIUS, ANICIUS MANLIUS TORQUATUS SEVERINUS (455/70–524/6). *De Syllogismo Categorico* and *Introductio ad Syllogismos Categoricos*. Paris, 1860. *Patrologiae Cursus Completus, Series Latina*, edited by J.-P. Migne. Vol. 64, cols. 761–832.

BOH, IVAN (1930–). 'Paul of Pergula on Suppositions and Consequences.' *Franciscan Studies*, 25 (1965), pp. 30–89.

BRADLEY, FRANCIS HERBERT (1846–1924). *Appearance and Reality: A Metaphysical Essay* (1893). 2nd ed., Oxford, Clarendon, 1897.

BURIDAN, JOHN (c. 1297–1358). *Summulae de Dialectica* (c. 1330). Edition entitled: *Compendium Totius Logicae*. Venice, 1499. Photoreprint, Frankfurt/Main, Minerva, 1965.

Sophismata (c. 1330). Edition entitled *John Buridan; Sophisms on Meaning and Truth*, translated and with an introduction by Theodore Kermit Scott. New York, Appleton–Century–Crofts, 1966.

BURLEY, WALTER (1274/5–1345). *De Obligationibus*. See GREEN, ROMUALD.

BUSCHERUS, HEIZO (or BUSCHER, 16th century). *De Ratione solvendi sophismata solide et perspicue ex P. Rami logica deducta et explicata, libri duo*. 3rd ed., Wittenberg, Hoffmann, 1594.

CAMPBELL, GEORGE (1719–96). *The Philosophy of Rhetoric* (1776), edited by Lloyd F. Bitzer. Carbondale, Southern Illinois University Press, 1963.

CARNAP, RUDOLF (1891–). *Introduction to Semantics*. Cambridge, Mass., Harvard University Press, 1942.

The Logical Syntax of Language (1934), translated by Amethe Smeaton. London, Routledge and Kegan Paul, 1937.

CASSIODORUS, SENATOR FLAVIUS MAGNUS AURELIUS (c. 487–580). *De Artibus ac disciplinis liberalium litterarum*. Paris, 1848.

Patrologia Cursus Completus Series Latina, edited by J.-P. Migne. Vol. 70, cols. 1149–1220.

CHASE, STUART (1888–   ). *Guides to Straight Thinking; with Thirteen Common Fallacies*. London, Phoenix House, 1959.

The Tyranny of Words. London, Methuen, 1938.

CHENU, MARIE DOMINIQUE (1895–   ). *Introduction à l'Etude de Saint Thomas d'Aquin*. Montréal, Institut d'Etudes Médiévales, 1950.

CICERO, MARCUS TULLIUS (106–43 B.C.). *De Divinatione*, with an English translation by W. A. Falconer. London, Loeb Classical Library, 1923.

De Natura Deorum, with an English translation by H. Rackham. London, Loeb Classical Library, 1933.

Topica, with an English translation by H. M. Hubbell. London, Loeb Classical Library, 1949.

COHEN, MORRIS RAPHAEL (1880–1947) AND NAGEL, ERNEST (1901–   ). *An Introduction to Logic and Scientific Method*. London, Routledge and Kegan Paul, 1934.

COLE, RICHARD (1926–   ). 'A Note on Informal Fallacies.' *Mind*, 74 (1965), pp. 432–3.

COPI, IRVING MARMER (1917–   ). *Introduction to Logic* (1953). 2nd ed., New York, Macmillan, 1961.

CUSACK, CYRIL AND OTHERS. 'The Cinema is the Highest Form of Art.' The text of the medieval disputation broadcast from Lincoln's Inn on 23 January 1950. *Blackfriars*, 31, 363 (1950), pp. 250–68.

DE MORGAN, AUGUSTUS (1806–71). *Formal Logic, or, The Calculus of Inference, Necessary and Probable* (1847). 2nd ed., edited by A. E. Taylor. London, Open Court, 1926.

A Budget of Paradoxes (1872), reprinted with the author's additions from the Athenaeum. 2nd ed., edited by D. E. Smith. New York, Dover Publications, 1954. 2 vols. in 1.

DESCARTES, RENE (1596–1650). *Correspondance*, with an introduction and notes by C. Adam and G. Milhaud. Paris, Presses Universitaires de France, 1936–63. 8 vols.

The Philosophical Works of Descartes, rendered into English by Elizabeth S. Haldane and G. R. T. Ross (1911–12). New York, Dover Publications, 1955. 2 vols. Vol. 1 contains *Passions of the Soul*.

DIODORUS SICULUS (Late 1st century B.C.). *The Library of History of Diodorus of Sicily*, with an English translation by C. H. Oldfather. London, Loeb Classical Library, 1933–67. 12 vols. Vol. 5 contains Book 12.

DIOGENES LAERTIUS (*c.* 230). *Lives of Eminent Philosophers*, with an English translation by R. D. Hicks. London, Loeb Classical Library, 1925. 2 vols.

D'ISRAELI, ISAAC (1766–1848). *Quarrels of Authors* (1814) in *The Calamities and Quarrels of Authors: with some inquiries respecting their moral and literary characters, and memoirs for our literary history*. A new edition, edited by his son, B. Disraeli. London, Routledge, Warne, 1863.

DUNCAN, WILLIAM (1717–60). *The Elements of Logick* in *The Preceptor: Containing a general course of education, wherein the first principles of polite learning are laid down in a way most suitable for trying the genius and advancing the instruction of youth*, by Robert Dodsley. London, J. Dodsley, 1748.

ENGEL, SRUL MORRIS (1939–). 'Hobbes's "Table of Absurdity".' *Philosophical Review* 60 (1961), pp. 533–43. Reprinted in *Hobbes Studies*, edited by K. C. Brown. Oxford, Blackwell, 1965. Also in Engel, S. Morris, *Language and Illustration: Studies in the History of Philosophy*. Amsterdam, Martinus Nijhoff, 1969.

FEARNSIDE, WILLIAM WARD (1913–) AND HOLTHER, WILLIAM BENJAMIN (1910–63). *Fallacy, the Counterfeit of Argument*. Englewood Cliffs, N. J., Prentice-Hall, 1959.

FRAUNCE, ABRAHAM (d. 1633). *The Lawiers Logike, Exemplifying the Praecepts of Logike by the Practise of the Common Lawe*. London, William How, 1588.

FREGE, GOTTLOB (1848–1925). *The Basic Laws of Arithmetic*, a translation by Montgomery Furth of selections from *Grundgesetze der Arithmetik* (1893). Berkeley, University of California Press, 1964.

'On Sense and Reference', a translation of 'Über Sinn und Bedeutung' (1892) in *Translations from the Philosophical Writings of Gottlob Frege*, by Peter Geach and Max Black (1952). 2nd ed., Oxford, Blackwell, 1960.

GALEN (CLAUDIUS GALENUS, 2nd century). *Opera, quae exstant*, edited by C. G. Kühn. Leipzig, Cnobloch, 1821–33. *Medicorum Graecorum*. 20 vols. Vol. 5 contains *Galeni de cuiusque animi peccatorum dignotione atque medela libellus*, pp. 58–103, and vol. 14, *Galeni de sophismatis seu captionibus penes dictionem*, pp. 582–98. Greek and Latin parallel texts.

GALILEI, GALILEO (1564–1642). *Dialogues Concerning Two New Sciences* (1638), translated by Henry Crew and Alfonzo de Salvio. New York, Dover Publications, 1914.

GARDNER, MARTIN (1914–). *Fads and Fallacies in the Name of Science* (formerly published (1952) under the title, *In the Name of Science*). New York, Dover Publications, 1957.

GASSENDI, PETRUS (1592–1655). *Opera Omnia*. Lyons, 1658. Facsimile edition: Stuttgart–Bad Cannstatt, Frommann, 1964. Vol. 1 contains *Syntagma Philosophicum*.

GAUTAMA (or GOTAMA, also AKṢAPĀDA, 2nd century?). *Nyāyasūtras* in *Gautama's Nyāyasūtras, with Vātsyāyana-Bhāsya*, translated, with notes, by Gaṅgānātha Jhā. Poona, Oriental Book Agency, 1939. 2 vols. Vol. 1 is Sanskrit, vol. 2 English translation.

GEACH, PETER THOMAS (1916–). *Reference and Generality: An Examination of Some Mediaeval and Modern Theories*. Ithaca, New York Cornell University Press, 1962.

GILBERT DE LA PORREE (GILBERTUS PORRETANUS, *c*. 1070–1154). *Liber de sex principiis*. Paris, 1855. *Pattologia Cursus Completus Series Latina*, edited by J.-P. Migne. Vol. 188, cols. 1257–70.

GILBY, THOMAS (1902–). *Barbara Celarent: a Description of Scholastic Dialectic*. London, Longmans, Green, 1949.

GRABMANN, MARTIN (1875–1949). *Die Introductiones in Logicam des Wilhelm von Shyreswood*, edited, and with an introduction by Martin Grabmann. *Sitzungsberichte der Bayerischen Akademie der Wissenschaften, Philosophisch-historische Abteilung*, 10 (1937), pp. 1–106.

'Die Sophismataliteratur des 12. und 13. Jahrhunderts mit Textausgabe eines Sophisma des Boetius von Dacien.' Münster, Aschendorff, 1940. *Beiträge zur Geschichte der Philosophie und Theologie des Mittelalters*, vol. 36, No. 1.

GREEN, ROMUALD (–). *An Introduction to the Logical Treatise 'De Obligationibus', with critical texts of William of Sherwood and Walter Burley*. Thesis presented at the Catholic University of Louvain, 1963. 2 vols.

GRENE, MARJORIE (GLICKSMAN, 1910–). *A Portrait of Aristotle*. London, Faber and Faber, 1963.

GRICE, HERBERT PAUL (–) AND STRAWSON, PETER FREDERICK (1919–). 'In Defense of a Dogma.' *Philosophical Review*, 65 (1956), pp. 141–58.

GROSSETESTE, ROBERT (1170–1253). *Die Philosophischen Werke des Robert Grosseteste, Bischofs von Lincoln, zum erstenmal vollständig in kritischer Ausgabe* by L. Baur. Münster i. Wien, Aschendorff, 1912. *Beiträge zur Geschichte der Philosophie des Mittelalters*, No. 9.

HAMBLIN, CHARLES LEONARD (1922–). *Elementary Formal Logic: a Programmed Course* (1966). London, Methuen, 1967.

HAMILTON, SIR WILLIAM (1788–1856). *Lectures on Metaphysics and Logic*, edited by H. L. Mansel and John Veitch. 3rd ed., Edinburgh, Blackwood, 1874. Vols. 3–4 contain *Lectures on Logic* (*c.* 1837–8).

HEAD, FREDERICK DEWAR (–). *Meetings: the Regulation of and Procedure at Meetings of Companies and Public Bodies and at Public Meetings* (1925). 6th ed., London, Pitman, 1957.

HEMPEL, CARL GUSTAV (1905–). 'Studies in the Logic of Confirmation.' *Mind*, 54 (1945), pp. 1–26, 97–121.

HERODOTUS (5th century B.C.). *The Histories*, translated and with an introduction by Aubrey de Sélincourt. Harmondsworth, Middlesex, Penguin Books, 1954.

HOBBES, THOMAS (1588–1679). *Opera Philosophica quae Latine Scripsit Omnia*, edited by William Molesworth. London, Bohn, 1839–45. 5 vols. Vol. 1 contains *Computatio Sive Logica* (1655).

HOWELL, WILBUR SAMUEL (1904–). *Logic and Rhetoric in England, 1500–1700* (1956). New York, Russell and Russell, 1961.

HUGHES, GEORGE EDWARD (1918–) AND CRESSWELL, MAXWELL JOHN (1939–). *An Introduction to Modal Logic*. London, Methuen, 1968.

HUME, DAVID (1711–76). *An Inquiry Concerning Human Understanding* (1748) *and other Essays*, edited and with an introduction by Ernest C. Mossner, New York, Washington Square Press, 1963.

A Treatise of Human Nature (1739), reprinted from the original edition in 3 volumes and edited with an analytical index by L. A. Selby-Bigge. Oxford, Clarendon, 1960.

ISAAC, JEAN (1910–). *Le Peri Hermeneias en Occident de Boèce à Saint Thomas: Histoire Littéraire d'un Traité d'Aristote*. Paris, Vrin, 1953.

JAEGER, WERNER WILHELM (1888–1961). *Aristotle: Fundamentals of the History of his Development* (1923), translated with the author's corrections and additions, by Richard Robinson. Oxford, Clarendon, 1934.

JOHN OF SALISBURY (BISHOP OF CHARTRES, 1115/20–80). *The Metalogicon of John of Salisbury, a Twelfth-Century Defense of the Verbal and Logical Arts of the Trivium* (1159), translated, with an introduction and notes by Daniel D. McGarry. Berkeley, University of California Press, 1962.

JOSEPH, HORACE WILLIAM BRINDLEY (1867–1943). *An Introduction to Logic* (1906). 2nd rev. ed., Oxford, Clarendon, 1916.

JUNGE, JOACHIM (JOACHIMUS JUNGIUS, 1587–1656). *Logica Hamburgensis, hoc est, institutiones logicae in usum Schol. Hamburg conscriptae* ... (1635). 2nd ed., Hamburg, Bartholdi Offermans, 1638.

BIBLIOGRAPHY 311

KAMIAT, ARNOLD HERMAN (–). *Critique of Poor Reason.* New York, privately printed (F. M. Rapp), 1936.

KENNY, COURTNEY STANHOPE (1847–1930). *Outlines of Criminal Law* (1902). 19th ed. rev. by J. W. C. Turner, Cambridge, University Press, 1966.

KEYNES, JOHN NEVILLE (1852–1949). *Studies and Exercises in Formal Logic; including a generalisation of logical processes in their application to complex inferences* (1884). 4th ed. rewritten and enlarged. London, Macmillan, 1906.

KNEALE, WILLIAM CALVERT (1906–) AND KNEALE, MARTHA (1909–). *The Development of Logic.* Oxford, Clarendon, 1962.

LAPLACE, PIERRE SIMON (SIMON, PIERRE, MARQUIS DE LA PLACE, 1749–1827). *Philosophical Essay on Probabilities* (1795), translated and edited by R. W. Truscott and F. L. Emory. New York, Dover Publications, 1951.

LEIBNIZ, GOTTFRIED WILHELM (1646–1716). *New Essays Concerning Human Understanding* (1765), translated, with notes, by A. G. Langley. La Salle, Illinois, Open Court, 1949.

LEVER, RALPH (d. 1585). *The Arte of Reason, rightly termed Witcraft: Teaching a Perfect Way to Argue and Dispute.* London, H. Bynneman, 1573.

LEVI, ANTHONY (1929–). *French Moralists, the Theory of the Passions,* 1585–1649. Oxford, Clarendon, 1964.

LOCKE, JOHN (1632–1704). *An Essay Concerning Human Understanding* (1690), edited, with an introduction, by John W. Yolton. London, Dent, 1961. 2 vols.

LUCIAN OF SAMOSATA (*c.* 130–80). *Works,* with an English translation by A. M. Harmon. London, Loeb Classical Library, 1915. Vol. 2 contains *Philosophies for Sale.*

ŁUKASIEWICZ, JAN (1878–1956). *Aristotle's Syllogistic, from the Standpoint of Modern Formal Logic* (1951). 2nd enl. ed., Oxford, Clarendon, 1957.

MATES, J. R. BENSON (1919–). *Stoic Logic* (1953). Berkeley, University of California Press, 1961.

MELANCHTHON, PHILIPP (or SCHWARZERD, 1497–1560). *Opera quae supersunt omnia,* edited by C. G. Bretschneider. Schwetschke, 1834–60. 28 vols. Vol. 12 contains *Erotemata Dialectices* (1528).

MELLONE, SYDNEY HERBERT (1869–1956). *An Introductory Text Book of Logic* (1902). 7th ed., Edinburgh, Blackwood, 1914.

MERCIER, CHARLES ARTHUR (1852–1919). *A New Logic.* London, Heinemann, 1912.

MILL, JOHN STUART (1806–73). *A System of Logic, Ratiocinative and Inductive, being a connected view of the principles of evidence and the methods of scientific investigation.* London, Longmans, Green, 1843.
    Utilitarianism (1863) in *Utilitarianism, Liberty and Representative Government.* London, Everyman, 1910.

MINIO-PALUELLO, LORENZO (1907–). 'The "Ars Disserendi" of Adam of Balsham Parvipontanus.' *Mediaeval and Renaissance Studies*, 3 (1954), pp. 116–69.
    'Iacobus Veneticus Grecus, Canonist and Translator of Aristotle.' *Traditio*, 8 (1952), pp. 265–304.

MONTAIGNE, MICHEL EYQUEM DE (1533–92). *Essays*, translated by John Florio (1605). London, Everyman, 1910. 3 vols. Vol. 3 contains *An Apologie of Raymond Sebond* (1580).

MOORE, GEORGE EDWARD (1873–1958). *Principia Ethica.* Cambridge, University Press, 1903.

MORE, SIR THOMAS, *saint* (1478–1535). *The Complete Works of Saint Thomas More*, edited by Edward Surtz and J. H. Hexter. New Haven, Yale University Press, 1965. Vol. 4 contains *Utopia* (1516).

MURRAY, RICHARD (–). *Murray's Compendium of Logic, with a corrected Latin text, an accurate translation and a familiar commentary*, by J. Walker. New edition, with explanatory notes from Whately, Mill, Gray, etc. Questions for exercise, a praxis, etc. London, 1847.

OESTERLE, JOHN ARTHUR (1912–). *Logic: The Art of Defining and Reasoning* (1952). 2nd ed., Englewood Cliffs, N. J., Prentice-Hall, 1963.

ONG, WALTER JACKSON (1912–). *Ramus: Method and the Decay of Dialogue: from the art of discourse to the art of reason.* Cambridge, Mass., Harvard University Press, 1958.

PAETOW, LOUIS JOHN (1880–1928). 'The Arts Course at Medieval Universities, with Special Reference to Grammar and Rhetoric.' *Illinois University Studies*, vol. 3, No. 7 (1910), pp. 497–624. Thesis, University of Pennsylvania, 1906.

PETER OF SPAIN (PETRUS HISPANUS, later POPE JOHN XX or XXI, 1210/20–77). *Summulae Logicales*, edited by I. M. Bocheński. Torino, Rome, Marietti, 1947.
    Tractatus Syncategorematum and Selected Anonymous Treatises, translated by Joseph P. Mullally. Milwaukee, Wis., Marquette University Press, 1964.
    Treatise on the Major Fallacies (*Tractatus Maiorum Fallaciarum*) MS. Clm. 14458 fol. 1ʳ–28ʳ. Bayerische Staatsbibliothek, München.

PLATO (427-347 B.C.). *The Dialogues of Plato, translated into English with analyses and introductions*, by B. Jowett (1871). 3rd ed., Oxford, University Press, 1892. 5 vols. Vol. 1 contains *Euthydemus* and *Phaedrus*, vol. 3 contains *Republic*, vol. 4 contains *Parmenides*.

POPKIN, RICHARD HENRY (1923-   ). *The History of Scepticism from Erasmus to Descartes* (1960). Rev. ed., Assen, Van Gorcum, 1964.

POPPER, SIR KARL RAIMUND (1902-   ). 'New Foundations for Logic.' *Mind*, 56 (1947), pp. 193-235.

POTTER, STEPHEN (1900-   ). *Some Notes on Lifemanship; With a Summary of Recent Researches in Gamesmanship*. Illustrated by Frank Wilson. London, Hart-Davis, 1950.

PRIOR, ARTHUR NORMAN (1914-   ). *Formal Logic* (1955). 2nd ed. Oxford, Clarendon, 1962.

Past, Present and Future. Oxford, Clarendon, 1967.

PRIOR, MARY LAURA (1922-   ) AND PRIOR, ARTHUR NORMAN (1914-   ). 'Erotetic Logic.' *Philosophical Review*, 64 (1955), pp. 43-59.

QUINE, WILLARD VAN ORMAN (1908-   ). *From a Logical Point of View: Nine Logico-Philosophical Essays* (1953). 2nd rev. ed., Cambridge, Mass., Harvard University Press, 1961.

Mathematical Logic. Cambridge, Mass., Harvard University Press, 1951.

Review of *Reference and Generality: An Examination of Some Medieval and Modern Theories*, by Peter Thomas Geach. Ithaca, Cornell University Press, 1962. *Philosophical Review*, 73 (1964), pp. 100-4.

RAMUS, PETRUS (PIERRE DE LA RAMEE, 1515-72). *Aristotelicae Animadversiones*, together with *Dialectica Institutiones* (or *Partitiones*). Facsimile of the first editions, Paris, 1543, with an introduction by W. Risse. Stuttgart-Bad Cannstatt, Frommann, 1964.

Dialectique (1555). Critical edition with notes and commentary by M. Dassonville, Genève, Librairie Droz, 1964.

REISCH, GREGOR (d. 1525). *Margarita Philosophica* (*Pearl of Philosophy*, 1496). Strassbourg, Johannes Schottus, 1504.

RESCHER, NICHOLAS (1928-   ). *The Development of Arabic Logic*. Pittsburgh, Pennsylvania, University of Pittsburgh Press, 1964.

RIJK, LAMBERTUS MARIE DE (1924-   ). *Logica Modernorum; a Contribution to the History of Early Terminist Logic*. Assen, Van Gorcum, 1962-7. 2 vols. in 3. Vol. 1 contains *Glosses on the Sophistical Refutations, Parvipontanus Fallacies, Summa of Sophistical Refutations*, and *Viennese Fallacies;* vol. 2, part 2, contains *London Fallacies, Munich Dialectica* (*Dialectica Monacensis*), and *Tractatus Anagnini*.

ROBERT DE MELUN (d. 1167). *Quaestiones de Divina Pagina (Questions on Holy Writ, c.* 1145) in *Oeuvres de Robert de Melun,* edited by Raymond M. Martin. Louvain, Spicilegium Sacrum Lovaniense, 1932. *Spicilegium Sacrum Lovaniense, Etudes et Documents,* fasc. 13.

ROBINSON, RICHARD (1902–   ). *Plato's Earlier Dialectic* (1941). 2nd ed., Oxford, Clarendon, 1953.

ROSS, SIR WILLIAM DAVID (1877–   ). *Aristotle* (1923). 5th ed., London, Methuen, 1953.

ROWE, WILLIAM LEONARD (1931–   ). 'The Fallacy of Composition.' *Mind*, 71 (1962), pp. 87–92.

RYLE, GILBERT (1900–   ). 'If, So and Because' in *Philosophical Analysis, a Collection of Essays,* edited by Max Black. Englewood Cliffs, N. J., Prentice-Hall, 1950.

Plato's Progress. Cambridge, University Press, 1966.

SALMON, WESLEY CHARLES (1925–   ). *Logic.* Englewood Cliffs, N. J., Prentice-Hall, 1963.

SAVONAROLA, GIROLAMO (1452–98). *Compendium Logicae* (1490). Florence, Bartolomeo de Libri, 1497.

SCHIPPER, EDITH WATSON (1909–   ) AND SCHUH, EDWARD WALTER (1922–   ). *A First Course in Modern Logic.* London, Routledge and Kegan Paul, 1960.

SCHOPENHAUER, ARTUR (1788–1860). 'The Art of Controversy' in his *Essays from the Parerga and Paralipomena,* translated by T. Bailey Saunders. London, Allen and Unwin, 1951.

SEXTUS EMPIRICUS (2nd–3rd century). *Works,* with a translation by R. G. Bury. London, Loeb Classical Library, 1933–49. 4 vols. Vol. 1 contains *Outlines of Pyrrhonism* and vol. 2 *Against the Logicians.*

SIDGWICK, ALFRED (1850–1943). *Fallacies, a View of Logic from the Practical Side.* New York, Appleton, 1884.

'The Localisation of Fallacy.' *Mind,* 7 (1882), pp. 55–64.

SIMON DE TOURNAI (fl. 1184–1200). *Les Disputationes de Simon de Tournai* (1201), collected by Joseph Warichez. Texte inédit. Louvain, Spicilegium Sacrum Lovaniense, 1932. *Spicilegium Sacrum Lovaniense, Etudes et Documents,* fasc. 12.

SMITH, SYDNEY (1771–1845). Review of *The Book of Fallacies* by Jeremy Bentham. *Edinburgh Review,* 42 (1825), pp. 367–89. Contains 'The Noodle's Oration'.

SOMMERS, FREDERIC TAMLER (1923–   ). 'The Ordinary Language Tree.' *Mind,* 68 (1959), pp. 160–85.

'Types and Ontology.' *Philosophical Review,* 72 (1963), pp. 327–63.

SOPHONIAS (?) (*c.* 1300). *Anonymi in Aristotelis Sophisticos Elenchos Paraphrasis*, edited by Michael Hayduck. Berlin, Reimer, 1884. *Commentaria in Aristotelem Graeca*. Vol. 23, part 4.

SPARKES, ALONZO CLIVE WILLIAM (1935–   ). 'Begging the Question.' *Journal of the History of Ideas*, 27 (1963), pp. 462–3.

STCHERBATSKY, FEDOR (1866–1942). *Buddhist Logic* (1930). New York, Dover Publications, 1962. 2 vols.

STEBBING, LIZZIE SUSAN (1885–1943). *A Modern Introduction to Logic* (1930). 2nd rev. ed., London, Methuen, 1933.

Thinking to Some Purpose. London, Pelican, 1939.

STEENBERGHEN, FERNAND VAN (1904–   ). *Aristotle in the West*, translated by Leonard Johnston. Louvain, Nauwelaerts, 1955.

THOMAS AQUINAS, *saint* (1224/5–74). *De Fallaciis ad Quosdam Nobiles Artistas* (*On Fallacies: For the Benefit of Some Gentlemen Students for an Arts Degree*, 1244–5?) in *Opuscula Philosophica*, edited by R. M. Spiazzi. Torino, Rome, Marietti, 1954. No. 43.

On the Power of God (Quaestiones Disputatae de Potentia Dei). Literally translated by the English Dominican Fathers. London, Burns, Oates and Washbourne, 1932–4. 3 vols.

Quaestiones Disputatae. 9th ed., Torino, Rome, Marietti, 1953. 2 vols. Vol. 1 contains *De Veritate*, edited by R. M. Spiazzi.

THOULESS, ROBERT HENRY (1894–   ). *Straight and Crooked Thinking*. London, Hodder and Stoughton, 1930.

VIDYĀBHĀṢAṆA, MAHĀMAHOPĀDHYĀYA SATIS CHANDRA (1870–1920). *A History of Indian Logic (Ancient, Mediaeval and Modern Schools)*. Calcutta, Calcutta University, 1921.

WADDINGTON, CHARLES TZAUNT (1819–1914). *Ramus (Pierre de la Ramée): sa vie, ses écrits et ses opinions*. Paris, Meyruéis, 1855.

WALLIS, JOHN (1616–1703). *Institutio Logicae, ad communes usus accommodata* (1686). 3rd ed., Oxford, Lichfield, 1702.

WATTS, ISAAC (1674–1748). *Logick, or the Right Use of Reason in the Enquiry after Truth, with a Variety of Rules to Guard Against Error, in the Affairs of Religion and Human Life, as well as in the Sciences*. London, printed for John Clark and Richard Hett, 1725.

WEILER, GERSHON (1926–   ). 'On Fritz Mauthner's Critique of Language.' *Mind*, 67 (1958), pp. 80–7.

WHATELY, RICHARD (1787–1863). 'Logic' in *Encyclopaedia Metropolitana: or, Universal Dictionary of Knowledge, on an original plan*... edited by Edward Smedley, Hugh James Rose and Henry John Rose. London, Fellowes and Rivington, 1817–45. Vol. 1, ch. 1.

Elements of Logic (1826). 9th ed. London, Longmans, 1848.

'Rhetoric' in *Encyclopaedia Metropolitana*, 1817–45. Vol. 1, ch. 2.

Elements of Rhetoric, Comprising an Analysis of the Laws of Moral Evidence and of Persuasion, with Rules for Argumentative Composition and Elocution (1828), edited by Douglas Ehninger. Carbondale. Southern Illinois University Press, 1963. Photo-offset reprint of the 7th edition.

WILLIAM OF SHERWOOD (or SHYRESWOOD, SYRWODE, etc. 1200/10–66/71). *De Obligationibus.* (Authorship doubtful.) *See* GREEN, ROMUALD.

William of Sherwood's Introduction to Logic, translated, with an introduction and notes by N. Kretzmann. Minneapolis, University of Minnesota Press, 1966.

WILSON, CURTIS ALAN (1921–). *William Heytesbury: Mediaeval Logic and the Rise of Mathematical Physics.* Madison, University of Wisconsin, 1960.

WILSON, THOMAS (1525–81). *The Rule of Reason, conteining the Arte of Logique set forth in English and newely corrected by Thomas Wilson: whereunto is added a table, for the ease of the reader* (1551). London, Grafton, 1552.

WITTGENSTEIN, LUDWIG (1889–1951). *Philosophical Investigations* with a translation by G. E. M. Anscombe. Oxford, Blackwell, 1953.

Preliminary Studies for the 'Philosophical Investigations', generally known as the Blue and the Brown Books (dictated 1933–4, 1934–5 respectively). Oxford, Blackwell, 1958.

Tractatus Logico-Philosophicus (1922), translated by C. K. Ogden. 2nd ed., London, Kegan Paul, 1947.

WRIGHT, GEORG HENRIK VON (1916–). *An Essay in Modal Logic.* Amsterdam, North-Holland Publishing Company, 1951.

INDEX

'Abd al-Masīḥ ibn Nā'imah, 103
Abelard, 107–9, 113–14, 124, 218, 304
Academy (of Plato), 52–5, 90
Accent, Fallacy of, 22–5, 62, 81–2, 99–100, 102, 118, 120, 138, 176, 206, 219
Accents, 22–3
Accident, Fallacy of, 26–8, 30, 62, 68, 72, 84–7, 91, 112, 121, 138–9, 149, 150–1, 153, 171, 205, 207–8
Adam of Balsham, 110–14, 119, 124, 304, 312
Adjunction principle, 220–1
ad rem, 42, 174
Affirming the Consequent, Fallacy of, 35–7, *and see* Consequent
Agricola (Roelof Huusman), 136–8, 148, 158, 304
Airay, Christopher, 139
Alberic of Paris, 114
Albert of Saxony, 130
Albert the Great, St, 11, 115–16, 162, 304
Aldrich, Henry, 32, 49, 91, 158, 163, 169–70, 195–9, 304
Alexander Neckham, 111, 304
Alexander of Aphrodisias, 97–103, 105, 115, 118, 304
Alexander the Great, 53, 177
Alexandria, 9, 89, 105
Alexinus, nicknamed Elenxinus, 90

Al-Ghazzālī, 103
Ambiguity, 19, 49, 111–12, 138, 141, 151, 193, 212–13, *and see* Equivocation; of middle term, 197–9; discoverable only *a posteriori*, 96–7, 297–9
Ammonius Hermiae, 105
Amphiboly, Fallacy of, 16–18, 62, 80–3, 99–102, 107, 112, 118–19, 138, 141, 205–6, 218
Analogy: argument by, 176, 227; Fallacy of False, 142
Andronicus of Rhodes, 54, 89
Apuleius, 204
Aquinas, St Thomas, 22, 115–16, 126, 307, 310, 315
Åqvist, Lennart, 216–17, 304
Arab writings, 10, 103, 177, 305, 313
Arguments: didactic, 54–5, 59–60, 66–7, 117, 161; dialectical (= examination), 55, 59–61, 63, 67, 117, 161; contentious (sophistical), 55, 59–60, 62–5, 67, 117, 161; persuasive force of, 70–3, 94, 241; hypothetical, 36, 233–4, 238, 245–6; probabilistic, 240, *and see* Probability; conditions of appraisal of, 234–8, 240, 245; extrinsic, 94, 140, 192, 237
Arguments *ad*: 31–2, 41–4, 94, 159–67, 170–1, 173–5; *baculum*, 41, 44, 156; *hominem*, 31, 41–2, 94, 142, 157, 160–4, 171, 173–5,

Arguments *ad cont.*
181, 205–6, 218, 224, 241; *ignorantiam*, 41, 43–4, 160, 162, 164, 281; *judicium*, 160–1, 163, 166–7, 174; *misericordiam*, 41, 43; *orationem*, 161–2; *passiones*, 41, 163–4, 171; *populum*, 41, 44, 164, 174–5; *verecundiam*, 31, 41–3, 94, 157, 159–60, 162–4, 166, 171, 173–5, 205–6, 218; *vertiginem*, 41, 162–3; *fidem*, 41, 163–4; *nauseam*, 41; *superstitionem*, 41, 166; *socordiam*, 41, 166; *rem*, 42, 174

Aristotle, 9–13, 20, 22–7, 31–3, 35–6, 38, 40–1, 43, 49–55, 59–90, 93–4, 97–8, 100–5, 107–19, 122–7, 131, 133, 136–40, 143–4, 146, 148, 151–2, 155, 158, 160–3, 166, 170, 175–9, 184, 186, 188, 190–5, 203, 205–6, 213, 215, 219, 229, 241, 246–7, 254, 279, 284, 304–5, 309–13, 315; *Categories*, 67, 104, 116; *Ethics*, 114; *Heavens*, 133; *Interpretation*, 67, 104–5, 116, 203, 310; *Metaphysics*, 79, 105, 116; *Physics*, 64, 79, 133; *Poetics*, 26; *Posterior Analytics*, 67, 76, 79, 85, 104, 109, 116–17; *Prior Analytics*, 52–3, 66–9, 73–6, 79–80, 86, 104, 116–17, 178, 181, 304; *Protrepticus*, 55; *Rhetoric*, 26, 38, 43, 52–3, 60–1, 66, 69–73, 79–80, 84, 94, 104, 152, 178; *Sophistical Refutations*, 9–11, 20, 23, 31, 35, 50–5, 59–67, 74, 78–88, 91, 97–8, 100, 102–4, 106, 109, 111–12, 114–19, 124, 127, 143, 152, 161–2, 179, 190, 213, 304–5, 313, 315; *Soul*, 114, 155; *Topics*, 11, 54–5, 59–60, 62, 65, 67–8, 73–4, 93–4, 104, 111, 114, 116–17, 122, 127, 178, 191, 304

Arnauld, Antoine, 29, 136, 148, 150–8, 305, *and see* Port Royal Logic; classification of fallacies, 150–1, 155–7

Arnim, H. von, 93, 305

Asclepius (of Alexandria), 105

Athens, 9, 53–4, 89, 154

Authority, arguments from, 34–5, 42–3, 146, 157, 218, 225–6, 250, 278, *and see* Arguments *ad verecundiam*

Averroes, 103, 305

Avicenna, 103

Babbling, aim of argument, 51, 63, 88, 118, 123, 139

Bacon, Francis, 10, 46–7, 136, 143–7, 154, 156, 158, 161, 175, 305

Bacon, Roger, 114–16, 121, 179, 305

Barbara celarent verse, 117

Bar-Hillel, Y., 21, 305

Bechmann, Friedemann, 305

Begging the Question, Fallacy of, 32–5, 49, 63, 67–8, 72–7, 101, 122, 125, 138–9, 141, 148–51, 166–7, 169–71, 181, 192, 205, 213–15, 225–8, 246–8, 251, 254, 271–2, 279; epistemological account of, 76–7, 151, 246–8; thesis that all valid reasoning commits, 35, 95, 122, 152, 180, 226–8, 246–8

Belnap, Nuel D., Jr., 216, 305

Bentham, Jeremy, 10, 42, 49, 165–9, 175, 284, 306, 314; classification of fallacies, 165–6

Bergh, S. van den, 103, 306

Bingham, Peregrine, 165

INDEX

Black, Max, 13–14, 16, 18, 31, 34, 39, 306, 308, 314
Blundeville, Thomas, 139
Bocheński, Innocentius, 66, 178, 187, 203–4, 306, 312
Boethius, 89, 103–9, 112–14, 181, 194, 208, 305–6, 310; Boethian classification of fallacies, 105–6, 112–13, 117, 122, 208, 211–12, 273
Boh, Ivan, 203, 306
Boole, George, 10, 136, 176, 203
Bossuet, 161
Bradley, F. H., 41, 306
Bradwardine, Thomas, 133
Brinton, Crane, 165
Buddhist: writings, 177, 184, 188; debates, 186–7, 260
Burden of Proof, 162, 170, 172–6, 274–5, 279, 294–5
Buridan, John, 86, 117, 120, 130–3, 306
Burley, Walter, 126–7, 306, 309
Buscher, Heizo, 10, 136, 142–3, 148, 306
Byzantine writings, 10, 97–8, 101–2, 105, 304, 315

Campbell, George, 164, 306
Caraka, 178
Carnap, R., 40, 220, 223, 286, 306
Cassiodorus, 104, 194, 306
Casuistry, Indian classification of, 182–4, 188–9
Catos, two historical, 105
Cause, 78–80, 152–3, 274, *and see* False Cause
Charles II and the Royal Society, 38–9
Chase, Stuart, 11, 307
Chenu, M. D., 126, 307
Chrysippus, 90, 92–3
Cicero, 17, 46, 93–4, 103, 108, 137, 140, 142, 307

Circular reasoning, 32–5, 171, 225, 229, 249, 271–2, 282, *and see* Begging the Question
Clinchers (Respondent's Failures), 183–6, 188–9
Cohen, M. R., and Nagel, E., 13, 19–20, 27, 30, 34–5, 41, 45, 197, 204, 307
Cole, Richard, 21, 307
Collard, John, 163
Commentaries on the *Sophistical Refutations*: Greek, 97–102, 304; Byzantine, 97, 101–2, 105; Latin, 114–15, 162, 304–5, 309, 313; Arabic, 103, 177, 305; modern, 85–6, 305
Commitment (dialectical), 257–8, 261, 263–79, 301; whole truth, 270–1
Complex Question, Fallacy of, *see* Many Questions
Composition, Fallacy of, 18–22, 62, 72, 80–1, 83–4, 99–100, 102, 112, 118, 120, 138, 140, 151, 171, 219
Consequent, Fallacy of, 35–7, 47, 63, 68, 72, 84–7, 122, 138, 149, 194, 205–7, 273
Constantinople, 10, 89, 103, 135
Contravalidity, 204
Controversy, Indian classification of, 182–3, 188–9
Copi, I. M., 13, 17, 19–20, 22, 24–5, 28–31, 34–9, 42–4, 176, 205, 307
Cresswell, M. J., 208, 310
Cusak, C., 126, 307

Debate, *see* Disputation
Definition, Fallacy of False, 142
De Morgan, Augustus, 13, 17, 24, 28, 30, 136, 143, 175–6, 227, 307
Demosthenes, 70, 80, 90

Denying the Antecedent, Fallacy of, 36
Descartes, 148, 155, 159, 307
Dialectic (= study of dialogue), 40, 59, 65–6, 69–71, 92, 175, 179, 253–303; formal and descriptive, 256, *and see* Disputation; Obligation
Dickoff, James, and James, Patricia, 150, 307
Diṅnāga, 188–9, 204
Diodorus Siculus, 26, 307
Diogenes Laertius, 89–91, 93–4, 97, 308
Disputation: Greek, 33, 39, 51, 54, 61–2, 126–7, 218, 276–80; medieval, 126–8, 260, *and see* Obligation; Buddhist, 186–8, 260; same as syllogism, 117; dangers of, 157
D'Israeli, Isaac, 38, 308
Distribution of Terms, 125, 195–7, *and see* Undistributed Middle
Division, Fallacy of, 18–22, 62, 72, 80–1, 83–4, 99–100, 102, 112, 118, 120, 138, 140, 151, 171, 219
Dubitatio, 128–9, 162
Dudman, V. H., 282
Dumbleton, John, 133
Dumont, Etienne-Louis, 165, 306
Duncan, William, 163, 308

Edmund of Abingdon, St, 114
'Electra', 90
Elisabeth of Bohemia, 155
Emotion, arguments appealing to, 154–8, 192, 250, *and see* Arguments *ad*
Emphasis, fallacies due to, 24–5, 275–6
Engel, S. Morris, 148, 308

Enthymemes (in Rhetoric), 71–3; spurious, 71–2
Equivocation, 182, 249, 251, 284–303; formal account of, 192–3, 195–9; Fallacy of, 13–16, 21, 45, 62, 80–2, 86, 98–102, 105–7, 112, 118–19, 124, 138, 153, 197–8, 205–6, 218–23, 284–303; Boethian Fallacy of, 105, 107–8
Etymology, Fallacy of, 142, *and see* Figure of Speech
Eubulides of Megara, 89–90

Fallacies Dependent on Language (*in dictione*), 13–26, 62, 68, 71–2, 80–4, 98–102, 118–21, 123, 131, 138, 141, 181, 205–6, 218–23, 254–5, 279, 284–303; Outside Language (*extra dictionem*), 13–14, 26–40, 62–3, 68, 118, 121–3, 138–9, 148–9, 205–7, 254; Common and Proper, 141, 148–9; Formal (Logical) and Material (Informal), 36, 169–71, 181, 193–5, 205–6; Formal, classification of, 44, 194–5, 199–205; impossibility of classifying, 13; *see also names of particular Fallacies*
Fallacy, definition of, 12, 48, 138, 146, 224, 253, *and see* Refutation
False Cause, Fallacy of, 37–8, *and see* Non-Cause
Falsification (Falsity, Fallacy), aim of argument, 51, 63, 118, 139
Fearnside, W. Ward, and Holther, William, B., 11, 17, 26, 41–2, 308
Figure of Speech, Fallacy of, 25–6, 29, 62, 65, 81–2, 99–102, 118, 121, 138, 206, 222–3

Five Terms, Fallacy of, 45
Florio, John, 147, 312
Form and Content, 194–5
Form of Expression, *see* Figure of Speech
Four Terms, Fallacy of, 44–5, 153, 197–9
Fraunce, Abraham, 14, 16, 34, 136, 139–42, 148, 161, 308; classification of fallacies, 140–2
Frege, G., 124, 136, 203, 207–8, 308
Furth, M., 124, 308
Futile Rejoinders (*jāti*), 183–5, 188–9

Galen, 98, 101, 103, 308
Galileo, 133–4, 151, 308
Gambler's Fallacy, 164–5
Gaṅgeśa, 189
Gardner, Martin, 48, 309
Gassendi, 91, 148, 152, 309
Gautama (Akśapeda), 178–81, 184, 186, 188, 309
Geach, Peter, 196, 308–9, 313
Generalization, Hasty, 29, 37, 46, *and see* Induction; Secundum Quid
Genetic Fallacy, 45
George of Trebizond, 135
Gilbert de la Porrée, 114, 309
Gilby, Thomas, 16, 25, 309
Glosses on the Sophistical Refutations, 114, 313
Gorgias, 26, 55
Grabmann, M., 130–1, 309
Grammar, study of, 26, 108, 113; ambiguity due to, 107, 111, 121, *and see* Figure of Speech
Granger, Thomas, 139
Greeley, Horace, 119
Green, Romuald, 126–8, 309
Grene, Marjorie, 53, 309

Grice, H. P., and Strawson, P. F., 290, 309
Grosseteste, Robert, 115, 309

Hamblin, C. L., 204, 309
Hamilton, Sir William, 196, 310
Harvey, William, 143
Hasty Generalization, 29, 37, 46, *and see* Induction; *Secundum Quid*
Hayduck, M., 98
Head, F. D., 244, 310
Hempel, C. G., 280, 310
Herodotus, 17, 310
Heytesbury, William, 133, 316
Hobbes, Thomas, 147–8, 159, 161, 308, 310
Holther, William, B., 11, 17, 26, 41–2, 308
Home, Henry, 163
Howell, W. S., 18, 139, 310
Hughes, G. E., and Cresswell, M. J., 208, 310
Hughes, Master (of Oxford), 114
Hume, David, 29, 37, 47, 146, 148, 155, 163, 310

Idols (in Bacon), 143–6, 154, 156; of Nation or Tribe, 144–6, 156; of Cave, 145–6; of Palace or Market Place, 145–6; of Theatre, 145–6
Ignoratio Elenchi, 31–2, *and see* Misconception of Refutation
Ignoring the Issue, Fallacy of, 31, *and see* Misconception of Refutation
Illicit Process (Illicit Major, Illicit Minor), Fallacy of, 44, 171, 197, 200–1
Implication, formal, 192, 204, 207, 220–3

INDEX

Indian: writings, 10, 177–89, 204, 229, 309, 315; classification of fallacies, 180–4, 188–9
Induction, 29, 46–7, 151, 153, 192, 225–6, 237, 249–50, 273, 278, 280–2
Invention and Judgment (Disposition), 62, 137–8, 141, 150
Irrelevant Conclusion, Fallacy of, 31, *and see* Misconception of Refutation
Isaac, J., 104, 113, 310
Isocrates, 70

Jaeger, W. W., 53, 310
James, Patricia, 150, 307
James of Venice, 109, 114, 312
Jhā, Gaṅgānātha, 178, 180, 309
Johannes de Alliaco, 131
John of Salisbury, 111, 310
Joseph, H. W. B., 13, 15, 17, 26, 38–9, 47, 310
Junge, Joachim, 148–9, 310

Kama sūtra, 178
Kamiat, A., 11, 311
Kant, 163
Kathāvatthu, 187
Kenny, C. S., 295, 311
Keynes, J. N., 33, 36, 206, 213, 311
Kilmington, Richard, 133
Kneale, W. C., and M., 91–2, 124, 196, 203, 311
Kretzmann, N., 116–17, 316

Laplace, 164–5, 218, 311
Larousse Dictionary, 161
Larrabee, H. A., 165, 306
Legal and moral concepts, 15, 292; legal procedure, 40, 94, 161, 244–5, 256, 274, 295

Leibniz, 136, 162–3, 311
Leśniewski, 195
Le Vayer, F. de la Mothe, 146–7
Lever, Ralph, 14, 139, 311
Levi, A., 155, 311
Liar, paradox of, 35, 90–1, 230, 301
Lifemanship, 52
Lincoln's Inn debates, 126, 260, 307
Literature, study of, 113, 136–7
Locke, John, 10, 41, 136, 148, 158–64, 241, 311
London Fallacies, 115, 313
Lucian, 90–1, 145, 311
Łukasiewicz, J., 67, 311
Luther, 137
Lyceum, 53–4, 89

Mansel, H. L., 32, 196, 304, 310
Many Questions, Fallacy of, 33, 38–40, 49, 63, 68, 72–4, 80, 103, 112, 123, 125, 138–9, 149, 205, 215–18, 262–3, 268–9
Marlowe, Christopher, 139
Marsais, C. C. du, 32
Mates, Benson, 91–2, 311
Meaning, dialectical theory of, 285–303
Megarians, 89–93, 148
Melanchthon, 137–8, 311
Mellone, S. H., 44, 311
Mercier, C. A., 18, 311
Merton College, 133
Metaphor, 26, 108, 141
Method, 60, 137, 150, *and see* Scientific Method
Michael Ephesius, 98, 304
Migne, J.-P., 62, 109, 161, 194, 305–7, 309
Mill, J. S., 11, 26, 29, 35, 46, 48–9, 136, 175–6, 226–7, 246–7, 250, 312; classification of fallacies, 48–9, 175–6

Minio-Paluello, L., 109–11, 304, 312
Misconception of Refutation (*Ignoratio Elenchi*), Fallacy of, 31–2, 41, 49, 63, 68, 80, 87–8, 105, 121–3, 138–9, 149, 151, 170–1, 173, 205, 208, 251, 273
Modal Logic, 207–12; of syllogism, 67, 117, 195; epistemic and deontic, 87, 208, 238–9
Modality, Fallacy of Different, 87, 106–7, 122, 211–12
Modus ponens, 35–6, 249, 299; *tollens*, 207, 237
Montaigne, 147, 312
Moore, G. E., 48, 312
More, Sir Thomas, 135–6, 312
Mullally, J. P., 312
multiplex, 98–102, 115, 118
Munich Dialectica, 29, 115, 313
Murray, Richard, 163, 312

Nagel, E., 13, 19–20, 27, 30, 34–5, 41, 45, 197, 204, 307
Naturalistic Fallacy, 48
Negativity, Fallacy of, 200
Newton, John, 139
Nicole, Pierre, 148
Non-Cause (as Cause), Fallacy of, 37–8, 47, 49, 63, 68, 72, 78–80, 123, 138–40, 149–50, 152, 176, 194, 205–7, 273, 279, 281
Nyāya sūtra, 10, 178–89; classification of fallacies, 180–3

Obligation, 123, 125–33, 136, 162, 218, 260–3, 276
Ockham, William of, 44, 131
Oesterle, J. A., 13–14, 20, 24, 27, 30–1, 34, 36, 38, 40, 312
'Old' and 'New' Logics, 104, 107, 113
Ong, Walter J., 137, 312

Onlookers' concepts, 242–6, 303
Opaque contexts, 207–8
Oxford, 10, 114–15, 133, 158, 169; Dictionary, 161

Paetow, L. J., 113, 312
Paradox: aim of argument, 51, 63, 118, 139; of Liar, 35, 90–1, 230, 301; of ravens, 280; of Self-Reference, 128, 130, 301–3; dialectical, 301–3
Paralogisms, 69, *and see* Syllogisms, invalid
Paris, 107, 110–11, 115–16, 139, 148
Part, Fallacy of Different, 87, 106, 122, 211; and Whole, Fallacy of, 18–22, 71–2, 84, 140
Parvipontanus school, 110; *Parvipontanus Fallacies*, 114, 313
Passions, *see* Emotion
Paternal dog, 27–8, 57–8, 86, 132
Pathetic Fallacy, 48
Paul of Pergula, 135, 203, 306
Paul of Venice, 135
Peter of Spain (Pope John XXI), 11, 22–3, 98, 100–1, 115–16, 120–3, 125, 131, 136–7, 196, 213, 312
Petitio principii, *see* Begging the Question
Petrus de Bognovilla, 131
Petrus de Insula, 131
Physical Science, sophisms on, 133–4
Places, *see* Topics
Plato, 27–8, 52–62, 65, 70, 132, 134, 144–5, 154–5, 276–7, 313–14; Platonism, 289; *Euthydemus*, 27, 55–9, 81–2, 132; *Meno*, 54; *Republic*, 54, 60, 145, 154; Academy of, 52–5, 90

Ploucquet, 163
Points of Order, 283–4, 295–6, 300–3
Popkin, R. H., 147, 313
Popper, Sir Karl, 207, 313
Porphyry, 104; tree of, 84
Port Royal Logic, 29, 46–7, 148, 150–63; classification of fallacies, 150–1, 155–7
Poste, Edward, 85–6, 305
Post hoc, ergo propter hoc, 37–8, 79–80, 152–3
Potter, Stephen, 52, 313
Pragmatics, 40, 286
Praśastapāda, 188
Presumption, 170–6, 294–5, *and see* Burden of Proof; *juris sed non de jure*, 295
Presupposition (of question), 38–40, 216–18, *and see* Many Questions
Prior, A. N., 195, 211, 217, 313
Prior, M. L., and A. N., 217, 313
Probability, 155, 164–5, 240, 250; dialectical, 179
Proof, 192, 248–9; Fallacy of, 33, 122, 213; Burden of, 162, 170, 172–6, 274–5, 279, 294–5
Propositions and sentences, 218–23
Protagoras, 55
Pseudo-Alexander, 98, 304
Pseudo-Augustine, 111
Punctuation, Fallacy of, 17–18, *and see* Amphiboly
Pyrrho, 94
Pyrrhus, King, 16–18, 119

Question and answer, 256–7, 260–3, 268–71; in Greek debates, 51, 54–5, 58, 61–8, 77, 89–90, 276–9
Question-Begging, *see* Begging the Question

Questions, logic of, 40, 110, 215–18; leading, 40, 61, 277; risky and safe, 216–17; biased and multiple, 262–3, 268–9, 277, *and see* Many Questions
Quine, W. V., 19–20, 196, 203, 207, 289–90, 313

Ralph Roister-Doister, 18
Ramus, 10, 136–40, 142–3, 148, 150, 158, 163, 312–13, 315
Raw Meat, 29–30, 121, 132
Reductio ad impossibile, 38, 78–9, 122, 185, 237, 274, 279; *ad falsum*, 78–9; *ad absurdum*, 185, 279
Referential opacity, 207–8
Refutation (*elenchus, redargutio*), 51, 63, 118, 122, 123, 139; weak and strong senses, 162; Aristotle's definition of, 87, 104–5, 122
Reisch, Gregor, 137, 313
Relations, logical, 203–5, 213–14; three-term, 214
Relatum, Fallacy of Different, 87, 106, 122, 211
Relevance, fallacies of, 31, *and see* Misconception of Refutation
Rescher, Nicholas, 98, 103, 313
Respondent's Failures (Clinchers), 183–6, 188–9
Retraction, 263–5, 267–8
Rhetoric, 26, 53, 61, 65, 70–3, 94, 113, 137, 153, 164, 168, 175, 178–9
Rhetorica ad Herennium, 108
Rijk, L. M. De, 29, 32, 104, 108–9, 114–15, 117, 119, 162, 304, 313
Road and the Chasm, 96, 298
Robert de Melun, 126, 314

Robinson, Richard, 60–1, 277, 310, 314
Roger Bacon, 114–16, 121, 179, 305
Rome, 9, 89
Roscelin, 108
Ross, Sir David, 53, 75, 213–14, 305, 314
Rowe, W. L., 20–1, 314
Russell, Bertrand, 136, 194, 203
Ryle, Gilbert, 62, 228, 314

Saccheri, 163
Salmon, Wesley C., 13, 46–7, 314
Sanderson, Robert, 139
Savonarola, 137, 314
Schipper, E. W., and Schuh, E. W., 13, 17–18, 24–5, 31–3, 37–8, 314
Schopenhauer, 10, 175, 314
Schuh, E. W., 13, 17–18, 24–5, 31–3, 37–8, 314
Scientific Method, 137, 150, 248, 280–2; fallacies of, 45–7
Scott, T. K., 306
Secundum Quid, Fallacy, of 27–31, 37, 46, 62, 68, 80, 86, 112, 121–2, 138–9, 149, 151, 182, 205, 208–13; in reverse, 30, 153
Self-reference, 128, 130, 301–3
Sense and reference, 124
Sextus Empiricus, 36, 91–7, 146–7, 182, 226, 246–7, 250, 296–8, 314
Siddhasena, 188
Sidgwick, Alfred, 10, 172–3, 175–6, 274, 314
Siger of Brabant, 131
Sign (Example), arguments by, 71–2, 176
Simon de Tournai, 126, 314
Smith, Samuel, 139

Smith, Sydney, 168, 314
Socrates, 20, 54–8, 85, 144, 276–7
Solecism, aim of argument, 51, 63, 80, 118, 123, 139
Sommers, Fred, 292, 314
Sophisms, *generally see* Fallacies; medieval (non-Aristotelian), 128–33, 136; and sophistical reasoning (in Aristotle), 50, 55, 62–3, 92, 116
Sophists, 20, 26, 55–8, 144
Sophonias (?), 98, 100, 315
'Sorites', 90–1
Spanish and Portuguese, 219
Sparkes, A. C. W., 32, 315
Special Pleading, Fallacy of, 25
Spinoza, 41
Statistical fallacies, 46, 164–5
Stcherbatsky, F., 184–5, 204, 315
Stebbing, Susan, 11, 203, 315
Steenberghen, F. van, 114, 116, 315
Stephen, Sir James, 274
Stoics, 36, 89–93, 148, 151, 214; classification of fallacies, 91–2
Stove, D. C., 230
Stratagems, 51–2, 64, 147, 173, 175
Strawson, P. F., 290, 309
Strode, Ralph, 135
Summa of Sophistical Refutations, 114, 118–19, 162, 313
Superfluity (Superfluous Premiss), Fallacy of, 79, 92, 141, 148, 214
Supposition, 104, 106, 116–17, 123–5, 130, 136
Swineshead, Richard, 133
Syllogisms, 36, 44–5, 67–8, 117, 125, 150, 193, 195–203; invalid, 44–5, 49, 69, 86, 148, 194–5, 197–203, 207; Indian, 178–80

Tātparya, 180
Tense-logic, 211–12, *and see* Time, Fallacy of Different
Testimony, Fallacy of False, 142, 156–7, 164; evaluation of, 164, 218, *and see* Authority
Theodoric, 103–4
Theophilus of Edessa, 103
'Third man' argument, 65
Thomas Aquinas, St, 22, 115–16, 126, 307, 310, 315
Thouless, R. H., 11, 315
Time, Fallacy of Different, 87, 106–7, 122, 211–12
Top-corner adverbs, 208–12
Topic points and Points of Order, 283–4, *and see* Points of Order
Topics (*loci*, Places, Commonplaces), 62, 71, 116, 137–8, 140–2, 148, 150, 158, 191, 255, *and see* Aristotle *Topics*; Cicero
Translation, source of ambiguity, 112; 'Translation' in Abelard (= 'Transumption' in Adam), 108, 113, 119

Udall, Nicholas, 18
Uddyotakara, 188–9
Undistributed Middle, Fallacy of, 44, 171, 197–200
Univocation, Boethian Fallacy of, 106, 108–9
Utopia, happy inhabitants of, 135–6

Vācaspati Miśra, 180
Vātsyāyana, 178–81, 184, 309
'Veiled Figure', 85–7, 90–1, 121, 132, 207–8
Vicious circle, 33
Vidyābhūṣaṇa, M. S. C., 177, 187, 315
Viennese Fallacies, 17, 114, 313

Waddington, C. T., 139, 315
Wallies, M., 98, 304
Wallis, John, 158, 169, 315
Walter Burley, 126–7, 306, 309
Watts, Isaac, 80, 163–4, 169, 315
Webster's Dictionary, 32
Weiler, Gershon, 95, 315
Wesley, John, 163
Whately, Richard, 34, 38, 42, 136, 168–75, 195–9, 205, 217, 228, 235, 312, 315; classification of fallacies, 169–71
Whole and Part, Fallacy of, 18–22, 71–2, 84, 140
'Wholly', logic of, 208–11
Wife-beating, 38–40, 216–17
William of Sherwood, 115–26, 128–9, 196, 215, 262, 309, 316
Wilson, Curtis, 133, 316
Wilson, Thomas, 18, 139, 148, 156, 316
Wittgenstein, L., 95, 242, 285, 301, 316
Wright, G. H. von, 87, 129, 235, 316

Yaḥyā ibn 'Adī, 98, 103